Arthur Jacob

The Designing and Construction of Storage Reservoirs

Arthur Jacob

The Designing and Construction of Storage Reservoirs

ISBN/EAN: 9783337736521

Printed in Europe, USA, Canada, Australia, Japan

Cover: Foto ©Andreas Hilbeck / pixelio.de

More available books at **www.hansebooks.com**

THE
DESIGNING AND CONSTRUCTION
OF
STORAGE RESERVOIRS.

BY

ARTHUR JACOB, B. A.,

LATE EXECUTIVE ENGINEER FOR IRRIGATION H. M.
BOMBAY SERVICE.

REVISED AND EXTENDED BY

E. SHERMAN GOULD, M. Am. Soc. C. E.,

CONSULTING ENGINEER TO THE SCRANTON GAS AND
WATER CO.

NEW YORK:
D. VAN NOSTRAND, PUBLISHER,
23 MURRAY AND 27 WARREN STREET.
1888.

EDITOR'S PREFACE.

In preparing a revised edition of "Jacob's Storage Reservoirs," it was considered best, in view of the standard character of the work, to present it in its original form, with all additional matter separate from the text.

All of the editorial work, therefore, in the present volume, appears in the shape of foot-notes, and the additional pages appended to the original discussion.

E. S. G.

Scranton, Pa., January, 1888.

THE DESIGNING AND CONSTRUCTION

OF

STORAGE RESERVOIRS.

Before entering upon such considerations as affect the selection of reservoir sites and their construction, a brief allusion to some of the most ancient works for impounding water may not be uninteresting. Of these the most prominent examples are undoubtedly to be found in Hindostan, where the magnitude and antiquity of the storage works cannot fail to arrest attention. These great works, surpassing in their immensity what are conventionally esteemed to be the wonders of the world, the production of other countries and nations, took their origin in the necessities of the people and the variableness of the climate of India, and were, in fact, great public works on which the welfare of the people mainly depended. The climate of India, although singularly uniform in some respects from year to year, is remarkably variable as regards the rain-fall; and in order to guard against the disasters of famine and sickness, inevitably attendant on a scanty monsoon, the native princes were wont to make such provisions as large resources and an almost unlimited power enabled them, in order to obviate the difficulty that they had to contend with.

The rain records of India for several years past show that a scarcity of rain is indicated by periods of about five years, or that every fifth or sixth year is marked by a scanty rainfall over certain districts. The recurrence of these periods is, of course, not very clearly marked, but still it is sufficiently so to warrant, with approximate correctness, the prediction of scarcity and famine; and such deplorable recurrence is, as all are aware, now reigning in India, and visiting with destruction, by sickness and hunger, some thousands whose sole dependence is upon a fair season of rain, and the successful maturing of their little crop of grain.

The natural expedient for guarding against the recurrence of these periodical calamities was evidently to be found in husbanding a scanty supply of rain-water for the purpose of irrigation, and this the people of India appear to have understood. They took advantage, in certain districts, of every nook and ravine, whether large or small, and converted them into storage reservoirs by throwing across banks of earth, or bunds, as they are termed, producing, in certain districts, such an elaborate and complete system of irrigation as can only be compared, for cost and completeness, to our railway system in England. Taking fourteen districts in the Madras Presidency, where tank irrigation was most generally relied upon, the records of the Indian Government show that there are no less than 43,000 irrigation reservoirs now in effective operation, and as many as 10,000 more that have fallen into disuse, making a total number of 53,000 storage works. The average length of embankment is found to be about half a mile, the extreme limit of the series being a dam of the immense length of 30 miles. This ancient reservoir, called the Poniary tank, is no longer in use, the cost of maintaining such a length of bank in adequate repair having probably been found disproportionate to the advantage derived from the supply. The work, embracing an area of storage of between 60 and 80 sq. miles, remains, however, as a record of what the Hindoos are capable of. To quote a second example, there is the Veranum reservoir, now in actual operation as a source of supply, and yielding a net revenue of no less than £11,450 per annum. The area of the tank is 35 sq. miles, and the storage is effected by a dam 12 miles in length. In order to bring the immensity of this system of storage works within the reach of statistical minds, it has been calculated that the embankments contain as much earth as would serve to encircle the globe with a belt of 6 ft. in thickness. To show that these are not singular examples, one other embankment of remarkable size may be alluded to. This embankment, of somewhat singular construction, was built on the island of Ceylon, and bears testimony that the Singalese monarchs were not behind their neighbors in public spirit or enterprise. The embankment was composed of huge blocks of stone strongly cemented together, and covered over with turf, a solid barrier of 15 miles in length, 100 ft. wide at base, sloping to a top width of 40 ft., and extending across the lower end of a spacious valley.

Thus it will appear that the practice of embanking across valleys, for the purpose of retaining the surface water, has for ages been in operation. There is no doubt that the disposal of some of the most remarkable works in India is not what it might, with advantage, have been; the fact remains, however, that the desired end was attained, and if the earthworks were disproportionately extensive, it was a source of satisfaction, at least, for the projectors to know that they cost, as a general rule little or nothing, the practice in those days being to press whatever labor was required, rendering in return nominal wages or none at all.

The two main questions that it is proposed to submit for consideration are, first, the selection of a reservoir site; and, secondly, the leading principles to be observed in the designing and construction of storage works.

The purpose or purposes for which the work may be required will, of course, affect materially the choice of a position, as well as the details of the structure itself; but certain general principles are available for our guidance in every case, after considering which, it is proposed to dwell upon such points as apply to the special purposes for which reservoirs may be constructed.

The first and most essential point for accurate determination by the engineer is, undoubtedly, the amount of rainfall, both maximum and minimum, that may be expected in the district under examination; and, having arrived at reliable data on this point, the next consideration will obviously be, what amount may be made available, due allowance having been made for evaporation and absorption. When we know that the annual depth of rainfall taken all over the world varies, according to the locality, between zero and 338 in. or 28 ft. deep (which excessive amount was on one occasion registered in the hill district of Western India), it will be obvious how little ground there will be for assumption, in the examination of any district hitherto unexplored, with regard to the question of its rainfall. In the examination of any given country, however, there are certain phenomena connected with the rainfall that will be found of almost invariable acceptation, and may with advantage be borne in mind.

The rainfall will, as a general rule, be greatest in those districts that are situated towards the point from which the prevailing winds blow. If Great Britain for instance be taken, the western districts will be found the most rainy. The very reverse, however, of this phenomena is noticed in the neighborhood of mountain ranges. If the wind prevails from one side rather than from the other, it is found that the greatest rainfall is on the leeward side of the range, and the probable solution of the matter is, that the air, highly charged with moisture, is carried up the hills by the wind until it comes into a cold region of the atmosphere. Condensation of the watery vapor immediately takes place, and the result is a fall of rain on the side of the mountain range remote from the prevailing wind.

To this cause may also be attributed the fact that the rainfall is always greatest in mountainous districts, while it by no means follows that elevated plains are more abundantly supplied with rain than land lying nearer to the sea level. The principles are remarkably exemplified in the southern part of the Bombay Presidency, where the author has had occasion to study the subject of rainfall. The Western Ghauts run parallel to the coast, rising to a height of 4,500 ft. above the sea, and from the western support of the great table-land of the Deccan, the mean elevation of which may be taken at 3,000 ft. In the rainy season the southwest monsoon, blowing from the sea, impinges against the Ghauts, and while passing onwards to the Deccan, parts with its moisture to the average annual amount of 254 in. On a spur of mountain that runs eastward, the pluviometers are found to register but 50 in.; and about 40 miles farther inland the rainfall is not more in some places than 15 in., which is considerably less than that registered in the lower-lying districts of the Presidency.*

---

*In regard to this matter, Mr. Fanning says in his "Water Supply Engineering":—"A study of some of our principal river valleys independently, reveals the fact that their rainfall gradually decreases from their outlets to their more elevated sources." And again: "The oft-made statement, 'that rain falls most abundantly on the high land,' is applicable, in the United States, to subordinate water-sheds only, and in rare instances."

In civilized countries like our own, much valuable information is usually available regarding the rainfall, if not applying actually to the district under examination, then probably to some neighboring districts, enjoying the same physical characteristics; but when any project of great importance is in contemplation, it will not be sufficient to take the returns of adjoining districts as accurate information of the rainfall at the exact locality fixed upon for the construction of the works. It will be necessary to establish rain-gauges at different points over the catchment basin of the valley from which it is intended to obtain the supply; and daily observations of these gauges must be taken for comparison with a series of simultaneous observations taken and recorded at the nearest station at which the rainfall has been regularly and carefully noted. It is evident that a comparison of the several observations taken over the area of water-shed with those registered at the permanent station, will convey a just estimate of the amount of maximum and minimum rainfall that may be relied upon.

The amount of rain falling upon the ground is not, however, the point to be determined, though it will aid considerably as a guide to the engineer. A considerable quantity of all the rainfall is either absorbed by the ground or evaporated before it reaches the point at which it can be made available for storage. Regarding, then, first, the question of absorption, it must be apparent that no two districts, unless they are exactly identical in soil, inclination of surface, and under similar circumstances of cultivation, can give on examination the same comparative result of rainfall and evaporation. If one district or unit of area be similar to the other in all respects but the surface inclination, that which has the greatest slope will, as a rule, give the largest percentage of water available for storage, because of course there will be less time for the rain to be absorbed. Again, the degree of cultivation will materially affect the result when two areas, otherwise precisely similar in their physical conformation, come to be compared one with the other, it being evident that an open and well-drained soil will be more favorable to the retention of water falling upon it than compact and impervious land. In every case the physical features of a district will each and every one of them force itself on the attention, as influencing the conclusion to be arrived at. If any general rule can be applied, it may be said that the greater the slope of the valley, the more rapidly it will throw surface water off; the more denuded the surface is of soil of any kind, the less will the escape of rain-water be retarded; and the more compact the rocks composing the geological structure of a district, the better will the circumstances be for impounding water. The volcanic rocks and those of the granite order will be as favorable as any that could be desired; while, on the other hand, porous rocks, such as the sandstones, chalk, etc., are too absorbent to offer the desired conditions for storage. It is not here asserted that all the water absorbed by porous rocks is necessarily intercepted from passing away to contribute to storage supply; much of it may be lost by evaporation and absorption by vegetables, but a considerable portion will often be found to contribute in the form of springs, if the disposition of the strata be favorable.

As a further source of loss, evaporation from the ground as well as from the surface of the reservoir, must be taken into consideration. The circumstances attending the latter source of loss will be considered further on, as this does not affect the question of how much of the total rainfall may be made available.*

The question how much water will be evaporated at any moment from the surface of land is one involved in considerable difficulty; and so many disturbing elements enter into the solution of the problem, that its accurate determination

---

*It is customary, in this country, when estimating the available areas of water-sheds, to deduct the areas of all lakes, ponds and open bodies of water, as the evaporation from them is supposed to cancel the rainfall upon them.

may be regarded as hardly possible of attainment. The hygrometric state of the ground's surface, the aspect of the sky, the amount of wind, and the temperature, will all, in their degree, exercise a sensible influence on the amount of water that the ground will give off from its surface; so that, in fact, it is doubtful whether any reliable and philosophically correct conclusions can be arrived at. The resultant facts from such experiments as have been carefully conducted afford, after all, the only data for the engineer to arrive at any general conclusion by; and for forming a rough estimate for the probable available rainfall of a district, the following proportions of available actual rainfall may be accepted as furnishing general data; but they are not meant to obviate the necessity of a careful and specific examination of the circumstances likely to affect the design of any particular work:

Steep surfaces of granite, gneiss, and slate	100
Moorland and hill pasture	60 to 80
Flat cultivated country	40 to 50
Chalk	0 to 0

In order to arrive at more specific, and truly reliable results, the engineer will have to make a series of accurate observations on the discharge of the stream or streams that carry away the rainfall of a district; and by doing so, and at the same time comparing the result with the amount of rain registered by the gauges —which should also, of course, be kept with accuracy in the locality under examination—an approximately true estimate of the available rainfall will be arrived at.

If there is time in the preparation of a project to make the necessary examination of a district, it is evident that the results will speak for themselves; and there will be no necessity to enter into abstract speculations concerning the theory of the influences affecting loss by either evaporation or absorption.

In proportioning the size of a storage reservoir to the area of the catchment basin, the engineer will, of course, in the first instance be guided by the requirements of the work. The object of the undertaking may be any one of the following:

To husband a scanty rainfall.

To check the injurious effect upon the country by floods.

To add to the discharge of a stream, by preventing the escape of the flood waters.

The amount of storage will always be part of an engineer's data in designing works. It will either be his object to store the whole of the water that the drainage area will afford, which will be the case in impounding water for irrigation, for example; or a certain fixed demand, governed by the want of a town or other requirements, will determine the amount of the rainfall that it will be necessary to retain for supply. In England the demand for water supply may be reckoned at from 150 to 180 days, depending on the amount and the constancy of the rainfall; as a rule, the six months' supply will be the safest to adopt. The following table extracted from Mr. Beardmore's work, shows the proportions that have been observed in designing some of the best constructed reservoirs. (See Table A.)*

From this it appears that the proportion between the amount stored and the total rainfall varies between one-half and one-fourth.

---

*There are a number of discrepancies in this table, but as the figures in the different columns are functions of each other, it would be very difficult, if not impossible, to say where the errors lie.

Table A.

Locality.	Height above Sea.		Drainage Area.
	ft.	ft.	sq. m.
Bann Reservoirs, 1837-38....	400 to	2800	5.15
Greenock, 1827-28, flat moor.	512 to	1000	7.68
Bute, 1826................	200 to	350	7.80
Glencorse, Pentland Hills..	734 to	1600	6.00
Belmont, 1843, moorland....	850 to	1600	2.81
" 1844, "	....	....	....
" 1845, "	....	....	....
" 1846, "	....	....	....
Rivington Pike, 1847........	800 to	1545	16.25
Longendale "	500 to	1800	....
Swineshaw "	500 to	1800	....
Turton and Entwistle, 1836	500 to	1300	3.18
" " 1837			....
Bolton Waterworks..........	800 to	1600	.80
Ashton " 1844.....	800		.59

TABLE A—*Continued.*

Total Discharge for the Year.	Discharge per Square Mile.	Representing Rainfall per Annum.	Registered Rainfall per Annum.	Reservoir Room per Square Mile.	Proportion of Storage to Available Rainfall.
c. ft. per m.	c. ft. per m.	in.	in.	c. ft. in. mill.	
1092.6	210.2	48.0	72.0	56.0	½
1416.6	197.7	41.0	60.0	38.0	2–5
819.0	105.0	23.9	45.4	......	......
600.0	100.0	22.3	37.0	7.66	3.10
630.4	224.3	50.7	63.4	26.8	¼
412.8	146.4	33.3	50.0	......	......
511.2	181.9	41.2	55.0	......	......
411.3	146.3	33.2	49.8	......	......
2880.0	176.7	40.0	55.5	29.6	⅓
......	......	49.5	55.5	......	......
......	......	37.0	49.3	......	......
576.7	181.3	41.0	46.2	31.43	⅓
548.2	172.3	39.0	48.2	......	......
100.2	125.2	32.7	......	25.6	1–3
40.7	65.5	15.5	40.0	21.0	4–7

The rule suggested by Professor Rankine "for estimating the available capacity required in a store reservoir, that founded upon taking into account the supply as well as the demand," is probably the best that can be adopted in designing waterworks for the supply of a town; "for example, 180 days of the excess of the daily demand above the least daily supply, as ascertained by gauging and computation in the manner above described." In order that a reservoir of the capacity "prescribed by the preceding rule may be efficient, it is essential that the least available annual rainfall of the gathering grounds should be sufficient to supply a year's demand for water." In calculating the capacity of a storage reservoir, the consideration of the surface evaporation must not be disregarded, especially when the works are designed for tropical or very dry climates. The amount of loss will in some cases be very considerable, for whatever depth of water be assumed to pass away into the air, it must be regarded as extending over the whole surface of the reservoir; or, in fact, the cubic quantity will be equal to the product of the depth evaporated away and the mean surface area of the reservoir as the water rises or falls throughout the year. Some have gone the length of asserting that the amount of evaporation from the surface of large and deep bodies of water is probably nothing at all, or at any rate, not worthy of consideration; whilst others assume a much larger amount of loss than appears to be supported by observation. The following extract from the article "Physical Geography," published by the Society for Promoting Useful Knowledge, expresses intelligibly the conditions that tend to promote evaporation:

"Other things being equal, evaporation is the more abundant the greater the warmth of the air above that of the evaporating body, and least of all when their temperature is the same. Neither does much take place whenever the atmosphere is more than 15 deg. colder than the surface upon which it acts. Winds powerfully promote evaporation, because they bring the air into continual as well as into closer and more violent contact with the surface acted upon, and also, in the case of liquids, increase by the agitation which they occasion, the number of points of contact between the atmosphere and the liquid.

"In the temperate zone, with a mean temperature of $52\frac{1}{4}$ deg., the annual evaporation has been found to be between 36 in. and 37 in. At Cumana, on the coast of South America (N. lat. $10\frac{1}{2}$), with a mean temperature of 81.86 deg., it was ascertained to be more than 100 in. in the course of the year; at Guadaloupe, in the West Indies, it has been observed to amount to 97 in. The degree of evaporation very much depends upon the difference between the quantity of vapor which the surrounding air is able to contain when saturated and the quantity which it actually contains. M. Humboldt found that in the torrid zone the quantity of vapor contained in the air is much nearer to the point of saturation than in the temperate zone. The evaporation within the tropics, and in hot weather in temperate zones, is on this account less than might have been supposed from the increase of temperature."

Thus it appears that evaporation, under highly favorable conditions, may take place to the extent of 9ft. in depth—an allowance that will demand careful consideration in designing storage works. In India, where, from the extreme dryness of the atmosphere, the evaporation is found to be considerable, the usual allowance made by engineers for evaporation from the surface of storage reservoirs is at the rate of $\frac{1}{2}$ in. of depth per diem for eight months in the year. Regarding the results that have been arrived at in Bombay, this allowance would appear to be about double what is necessary, for the observations extending over five years give a mean daily evaporation of less than $\frac{1}{4}$ in. In Bombay, however, the atmosphere is much more humid than that experienced on the great tableland of the Deccan; and in Madras, where reservoirs are the specialty, it is probable that the actual loss is not far from being a mean between the two fractions. In Great Britain the mean daily evaporation is found to average less than the tenth of an inch.*

In estimating the quantity of storage water that will result from the drainage of any particular district, it will be essential to consider carefully the geological disposition of the strata characterizing the locality in which it is contemplated to establish the works. This, although a matter that may influence the effectiveness of an undertaking to the extent of success or failure, will appear to the purely practical man to imply a degree of refinement that is uncalled for. There will be no difficulty, however, in showing that the geological conformation of a district may be such as, on the one hand, to materially contribute to the efficiency of a

---

*See an interesting paper upon Evaporation, by Mr. Desmond FitzGerald, M. Am. Soc. C. E., contained in the Transactions of the American Society of Civil Engineers, for Sept., 1886.

storage reservoir, or on the other to prove
so defective that no engineering skill or
pecuniary outlay could remedy it. A
condition of geological structure, perhaps
the most favorable that could be imagined,
is that shown in Fig. 1. This diagram
represents a geological section taken at
right angles, or nearly so, to the axis of
the valley that it is proposed to convert to
the purpose of storage. This somewhat
peculiar structure is what is geologically
termed synclinal, the beds inclining away
from the axis of the valley, and is the result of an upheaving force having taken
place underneath the points of greatest
elevation. Subsequent to the upheaval
and consequent displacement of the
strata, the process of denudation has
taken place, cutting the upper beds, and
leaving the outcrop exposed, not only inside the basin, but in the adjoining
valleys at O and O. Now, it is evident
that if the highest ridges bounding the
valley be taken to mark the line of watershed, and therefore limiting the area of
the catchment basin, it is possible that

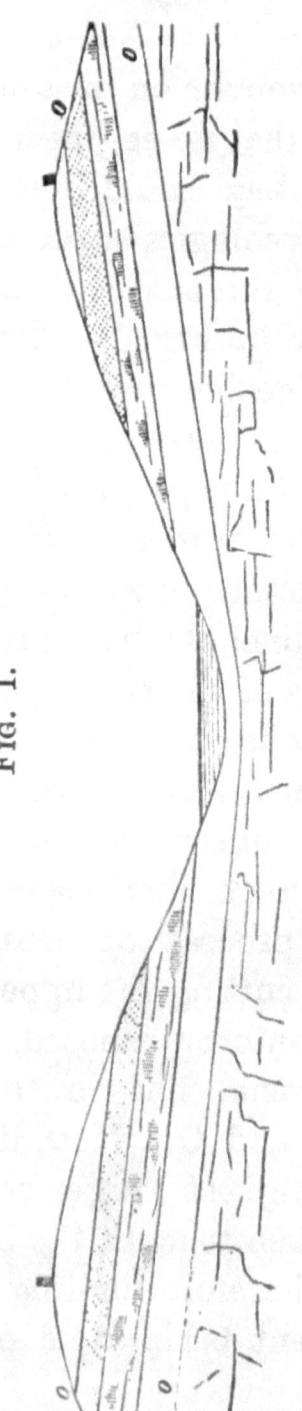

Fig. 1.

SECTION ACROSS SYNCLINAL AXIS.

the estimate of the amount of supply may be found far short of what the district will yield. A certain proportion of rain falling upon the outcrop at the points O O will be absorbed by such of the strata as are porous, and the water, percolating through the bedding, till an impervious stratum is met with, will find its way down the course of the stratification, till it ultimately reaches the reservoir in the form of springs, and contributes more or less to the maintenance of the supply. The converse of this condition of things will be readily understood by reference to Fig. 2. It also represents a section taken directly across the valley of the proposed reservoir. Here the strata of the earth's crust incline against each other consequent upon some disturbing force having taken place to elevate them, and are said to be anticlinal to the axis of the valley. In order to account for the formation of a valley on the summit of the ridge, that at first was thrown up, it is to be understood that the upper beds suffered fracture in the process of upheaval

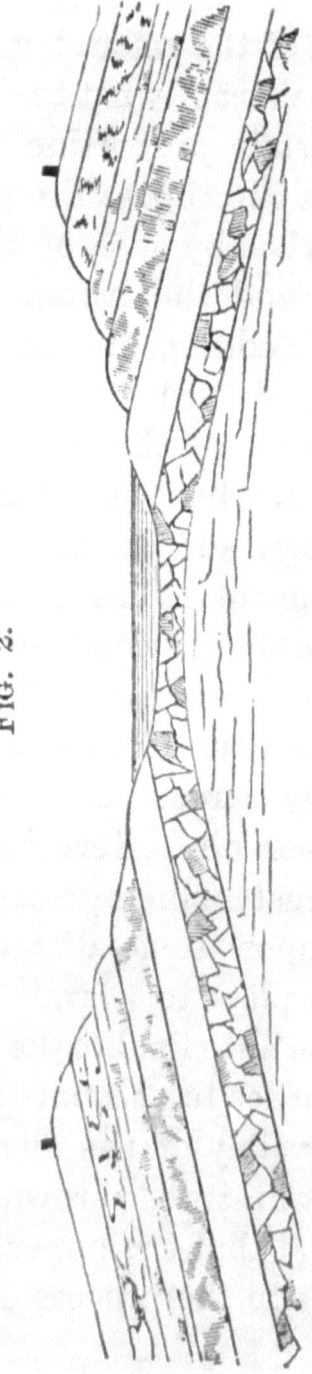

Fig. 2.

SECTION ACROSS ANTICLINAL AXIS.

and subsequently were exposed to denudation. These valleys of elevation are evidently not to be desired as situations for the establishment of storage reservoirs. The area of the gathering grounds will be much more limited than the extent of the water-shed would appear to indicate; and cannot safely be relied upon to give an estimate of the quantity of water that the valley will afford. A certain amount of water will undoubtedly pass over the surface in times of heavy and continued rain, before it can be absorbed; but there is no doubt that of all the water absorbed by the ground, by far the greater portion will follow the inclination of the strata, and come out as springs in the adjoining valleys.

Fig. 3 shows a geological section that combines in it favorable and unfavorable conditions for the storage of water. On one side the outcrops of the strata are found to extend beyond the highest point of water-shed line, whilst on the other side the strata incline away, producing such a condition as would favor the escape from the valley of the water absorbed.

Certain rules are in general use for estimating the quantity of the total rainfall that will be lost by absorption and evaporation, with a view to determining the proper proportion to be observed between the reservoir and the area of the catchment basin. Two-thirds of the whole fall is sometimes taken to represent the loss that may be expected from the drainage of any district; in general terms, one-third being assumed as the amount that may actually be intercepted for utilization. Some authors leave a much smaller margin, and state that fully two-thirds of the total rainfall may fairly be taken as available for storage. This is a large discrepancy when the application of the rules is taken to be general; but when the statements are applied to separate districts and different countries, there is nothing irreconcilable in them. General rules are undoubtedly of much value if they be received with qualification,

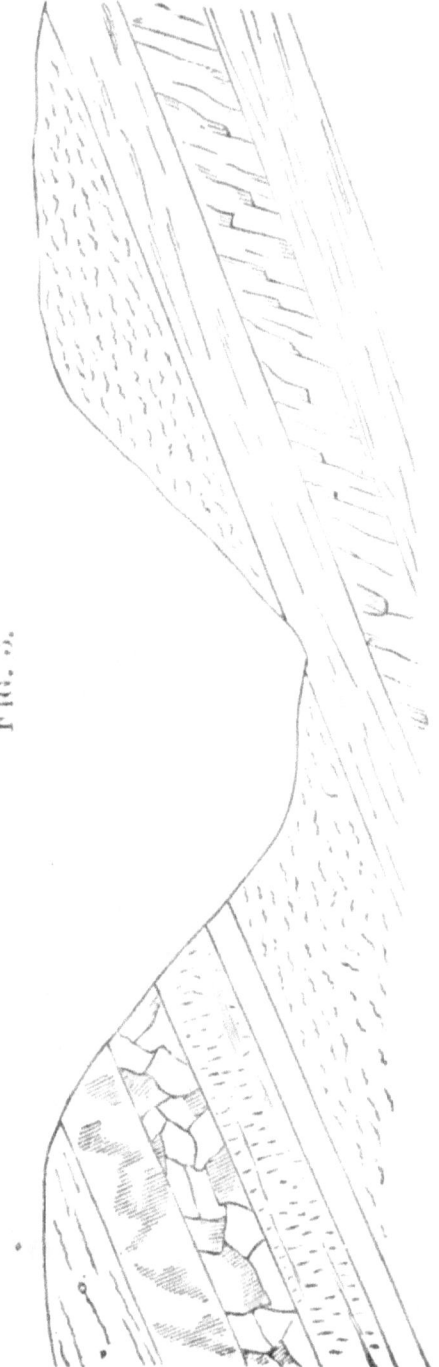

Fig. 3.

VALLEY OF DENUDATION.

and are not adopted as of absolutely universal application. They cannot, however, with safety be substituted for specific investigations, when so much depends on starting with accurate data.

RESERVOIR SITES.

The special requirements of each particular case will, as a general rule, go far towards determining the selection of a site for the establishment of storage works. Assuming, however, that there is a considerable extent of country situated advantageously in relative position to the locality at which it is proposed to utilize the water, and that there is a choice of ground, the point to be considered chiefly will be the natural lie of the country. To throw an embankment across a valley at any point without due regard to the configuration of the ground would most probably result in an expensive and ill-designed scheme; for under such circumstances the cost of the dam would bear a very large proportion to the quantity of water stored. It will rarely happen that, in the examination of the resources of any particular piece of country, some special features will not present themselves, favorable to the situation of storage works. The most advantageous disposition of the ground will be when two spurs of high land approach each other, forming a narrow outlet for the stream, and leaving a wide space above them in the valley for storage. Such a configuration is not uncommonly met with at the junction of two streams, as shown in Fig. 4. This is merely a sketch from memory, by the author, of a reservoir that he had designed in India for purposes of irrigation; and it will be evident that the disposition of the ground was singularly favorable in every respect for the construction of a large storage work. The area of the reservoir, as designed, was about three square miles, and the maximum depth 90 ft., the area of the catchment basin being about 60 square miles. Such favorable situations for storage are of somewhat rare occurrence ; for when the contour of the land is what is desirable, it may be that the area of water-shed is not adequate, or

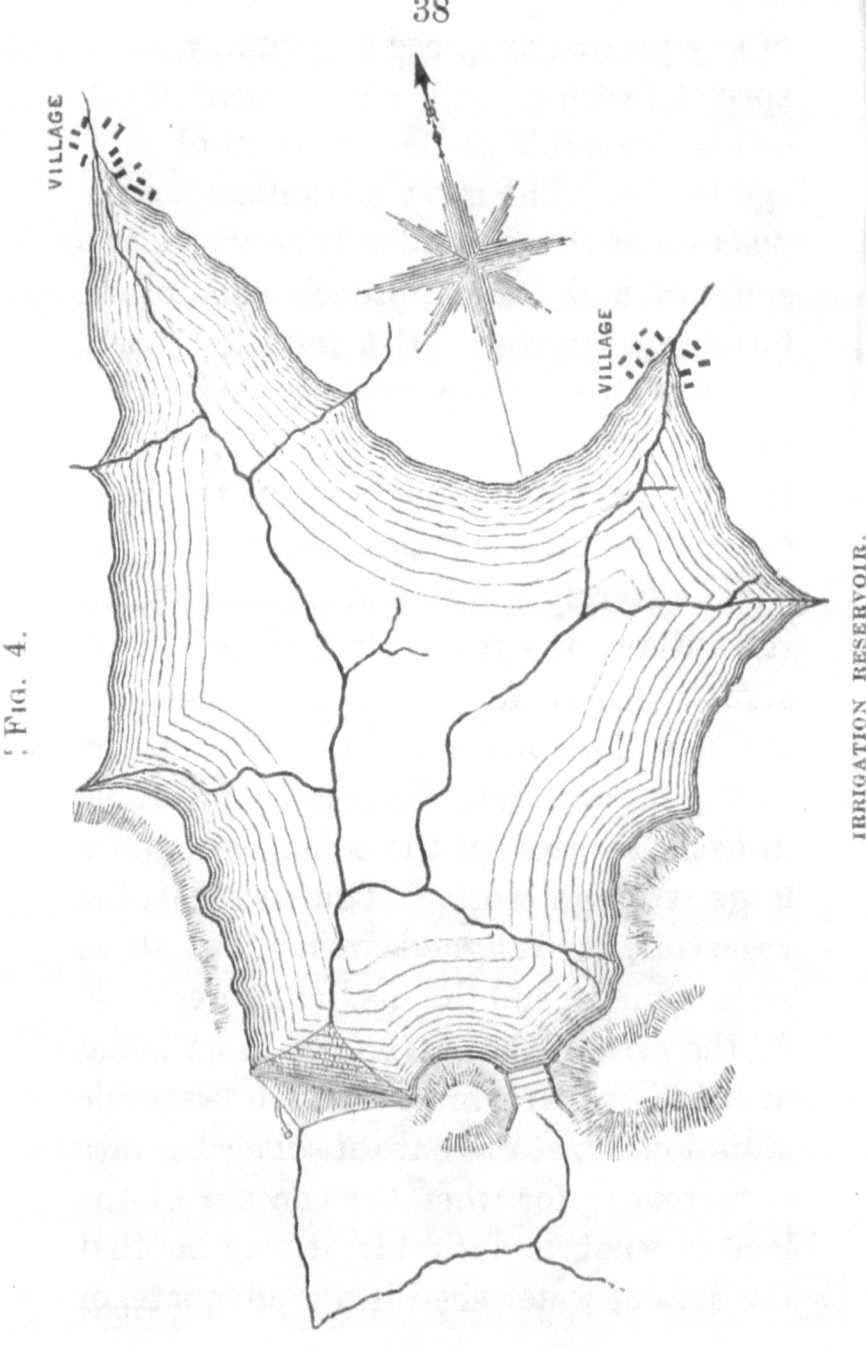

Fig. 4.

possibly the geological condition of the ground may be unfavorable, or the materials for the construction of a sound bank are not available. In examining large tracts of country in India, with a view to the establishment of irrigation reservoirs, the author found that more reliance was to be placed on a careful examination of the map in the first instance, than on the common plan of making personal explorations of the country. A good map will show at a glance, especially if the hill-shading has been carefully engraved, the points at which the supply will be found sufficient to justify the undertaking; and will probably furnish a pretty true indication of sites at which embankments may be advantageously constructed.

In tropical climates, where the rainfall is in places very scanty, and where the land is not of great value, it not unfrequently happens that such situations prove available for the establishment of large storage works as would not under any circumstances be made available in

England. These sites are to be found, not at the head of a valley, but at some considerable distance down the course of a stream, where, the general inclination of the country being slight, a low embankment serves to store a very large area of water. The apparent disadvantages of such a site for storage are the large area of land swamped and lost to the cultivator and to Government, and the great surface exposed to evaporation under a tropical sun and the influence of a dry wind. In India, the first objection is one of comparatively little moment, considering that in those districts where irrigation is most required the value of land is very trifling. From 1s. to 2s. is about an average rent per acre, where land is under dry crops; but when water is available, the cultivators can, with profit, afford to pay 30s. per acre. It is therefore evident that, so far as Government is concerned, there is no sacrifice in the matter, but, on the contrary, an unspeakable benefit is conferred on those land owners who hold farms below the reservoir; and an ample supply of water is stored in the dryest seasons to mature those crops whose failure almost inevitably reduces the people to the verge of starvation. The evaporation from these lakes is, beyond question, a source of very considerable loss, and one that admits of no possible abatement. Estimated as above, at about half an inch vertical for eight months of the year, the loss frequently amounts to one-third of the whole body of water stored. As a set-off against this and other objections, the facilities for constructing these reservoirs of great extent, are considerable. In the first place, the embankments, being very low, are rapidly and cheaply constructed by native workmen; and when finished, the head of water even at the deepest point is not sufficient to try the work to any great extent. Further, the greater the extent of the reservoir, the less inconvenience is experienced from silting. The streams, owing to the suddenness of the rainfall, come down heavily charged with earth in suspension, the mass of which is deposited like a miniature delta at the influx of the reservoir, instead of passing on and resting near the embankment, as invariably occurs in reservoirs of small extent. The immense consumption of water necessary to confer any appreciable benefit by irrigation is of itself the strongest argument in favor of these broad and shallow reservoirs; for it is not possible to find in the upper part of a valley such sites as would store the requisite quantity of water without an embankment of excessive dimensions; and moreover, the catchment area in such situations is not usually sufficient to serve, with a scanty rainfall, for the supply of a very large reservoir. It is not, of course, maintained that this mode of storing water is by any means applicable in England, for the circumstances and requirements in each case are wholly dissimilar.

SUPPLY.

The reservoir site being supposed everything that could be desired, as regards the disposition of the ground, the supply will next engage attention as a matter of course. Assuming that the gathering grounds are sufficiently extensive, it is presumed that the reservoir will be constructed to contain sufficient water to meet the maximum demand, whatever that may be calculated at; and in order to determine with accuracy what capacity the reservoir will have with different heights of embankment, it will be necessary to carry out certain leveling operations over the ground. The least elaborate manner of proceeding will be to run a series of cross-levels through the valley, referring all to the same datum, and by comparing these levels to ascertain what the average depth will be for a given height of bank. Having decided the height of the water-level, the next operation will be to contour round the basin, and to survey the boundary-line. In this way may be acquired sufficient knowledge as to the storage capacity, to justify the procedure with the work. When the execution of the project has been determined upon, it will be advisable to make a more accurate survey of the bed of the valley, and this can best be done by covering the whole plan with a series of contour lines at a vertical distance from each other of about 5 ft. This kind of survey will be of lasting value to the engineer, for it will enable him to calculate what quantity of water the reservoir will contain at each foot of depth; and, consequently, he will know, from a mere inspection of the gauge in the reservoir, how much water he has at his disposal for service.

It has been assumed that the gathering grounds are sufficient to maintain the requisite supply in the reservoir; but it may be well to pause and inquire what extent of water-shed will be sufficient to furnish a given supply, and what method may be adopted for supplementing an insufficient drainage area. It has before been remarked that the only reliable information, when there is any question as to the sufficiency of the rainfall or the area of the catchment basin, can be derived from careful gaugings of the stream or streams that may be depended upon to contribute to the supply. If the catchment area is very large as compared to the capacity of the reservoir, a mere inspection of the map and an exploration of the ground will generally be conclusive as to the sufficiency of the supply for storage. Should there not be such conclusive evidence on this point, it must be determined by measuring the quantity of water that absolutely flows off the ground, at the same time gauging the rainfall. This latter precaution would appear unnecessary, but in truth it is of great value, for it will furnish by comparison with the rainfall registers that have been kept through the same year, and a series of previous years, evidence as to the amount of available rainfall that may be expected during terms of comparative drought. If the supply of a town with water be the desideratum, the rule to be rigidly observed is that of making a minimum supply meet the maximum demand, and therefore it is of the highest importance to determine beyond any doubt, what the minimum yield of a catchment basin will be.

As a mode of supplementing an insufficiently large drainage area, catchment drains or feeders have frequently rendered good service. These are cuts that are carried outside the water-shed line to arrest the surface drainage and catch the contributions of small streams, and conduct the water into the reservoir. The greater the area enclosed between catchment drains and the water-shed line, the more valuable will they be as aids to the supply of the reservoir. They of course virtually extend the area of the catchment, adding so many square miles or acres to the rainfall.

DESIGNING OF WORKS.

Knowing the exact requirement of a given population, or rather having fixed, after every consideration, the daily consumption of every individual that it is proposed to supply, there will be no difficulty whatever in proportioning the reservoir to the demand upon it. It is sometimes necessary, however, to provide reservoirs for the purpose of preventing damage to the country by floods, and in this way the inconvenience and injury naturally consequent upon very sudden and excessive falls of rain may be to a great extent obviated. The duty of the reservoir will be to arrest all water in excess of what the stream can carry within its banks, and to dispose of this excess water, so to speak, in detail, after the excessive rainfall has become moderated. A comparison of a stream's discharge, taken at highest floods, with the quantity that it can carry without overflowing its banks, will show the excess that has to be retained by the reservoir; and these data can only be arrived at through a carefully kept record of the extent of the floods and of their duration. The maximum flood in this consideration will not be that which rises to the greatest height for a short time, but will be the product of the excess above what the river can discharge by the length of time the flood lasts; which will, in fact, be the necessary capacity of the reservoir.

The table given on a preceding page will afford an interesting study when compared with the following table, extracted in part from the same work. The first gives a comparative view of the volume of water gauged and stored in small hill districts, the last column indicating the proportion of the total available rainfall to the amount actually intercepted for storage. The following table shows the ordinary summer discharge of various rivers, streams, and springs, as unaffected by immediate rain. (See Table B.)*

Where the reservoir is designed to check the injurious effects of floods, the proportion of the storage to the rainfall will, in most cases, be much smaller than what would be necessary to provide for the better part of a whole year's fall of rain, for it is not probable that the maximum known flood can ever exceed the amount that it would be necessary to store for economic purposes.

---

*This table also contains several evident inaccuracies; where the point of error seemed apparent, correction has been made in the present edition.

TABLE B.

RIVERS.	Height above Sea.	
	Valley ft.	Hill. ft.
Thames at Staines, chalk, greensand Oxford clay, oolites, etc............	40 to	700
Severn at Stonebench, silurian......	400 to	2600
Loddon (February, 1850), greensand.	110 to	700
Nene, at Peterborough, oolites, Oxford clay and lias..................	10 to	600
Mimram, at Panshanger, chalk.......	200 to	500
Lee, at Lee Bridge, chalk (Rennie, April, 1796).......................	30 to	600
Wandle, below Carshalton, chalk....	70 to	350
Medway, dryest seasons (Rennie, 1787), clay.	....	
Ditto, ordinary summer run (Rennie, 1787)................................	....	
Verulam, at Bushey Hall, chalk......	150 to	500
Gade, at Hunton Bridge, chalk......	150 to	500
Plym, at Sheepstor, granite..........	800 to	1500
Woodhead Tunnel, millstone, grit....	1000	
Glencorse Burn......................	750 to	1600

Table B.—*Continued.*

Drainage Area.	Total Discharge for the Year.	Discharge per Sq. Mile.	Representing Rainfall per Annum.	Total average Rainfall per Annum.
sq. miles.	c. ft. per. min.	c. ft. per min.	in.	in.
3086	40,000	12.98	2.93	24.5
3900	33,111	8.49	1.98	....
221.8	3 000	13.53	3.01	25.4
620.0	5,000	8.45	1.88	23.1
50.0	1,200	24.3	5.5	26.6
570.0	8 880	15.58	3.53	....
41.0	1,800	43.9	9.93	24.0
481.5	2,209	4.59	1.04	....
481.5	2,520	5.23	1.19	....
120.8	1,800	14.9	3.37	....
69.5	2,500	36.2	8.19	....
7.6	500	71.4	15.10	45.0
....	139	....	....	46.0
6.0	130	21.6	4.9	37.4

PROPORTIONS OF BANK.

The proper proportion to be given to an embankment for the support of water is a question that appears to admit of a good deal of difference of opinion, some designers taking one view, some another, of the proper theory that is to determine the dimensions of a bank. Some few, with whom the author cannot agree on this point, maintain that a bank ought to be designed with strict reference to its theoretical power of resisting hydrostatic pressure, or the effort of the water to displace it. Regarding the question in its abstract form, it will be evident that any structure intended to sustain the pressure of water may be supposed to fail in one of two ways—either, in the first place, by yielding to the horizontal pressure of the water and overturning, or by progressive motion, *i. e.*, sliding on its base. In considering the first theory, that of resistance of overturning, the easiest method of examining the question will be to take a simple example of a vertical rectangular wall, and ascertain what power it exercises to resist the pressure of water. The pressure of water upon any plane surface immersed is known to be equal to the area of that surface, multiplied by the depth of its center of gravity below the level of the water. Generally speaking, the unit adopted in calculation is a foot; and the

FIG. 5.

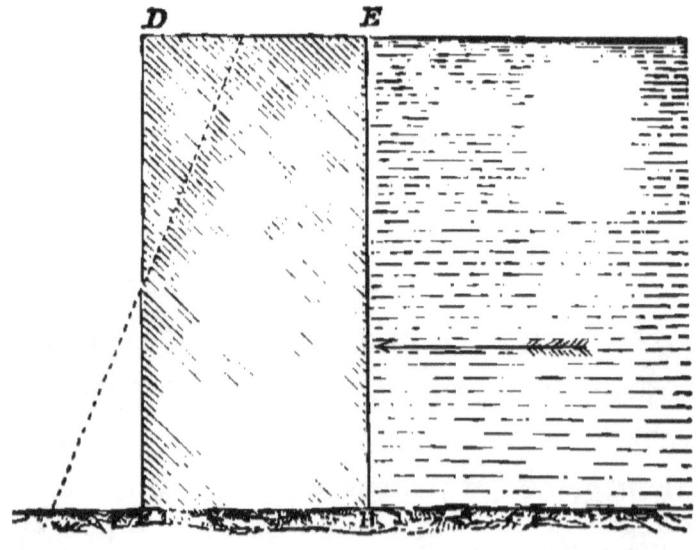

unit of water being taken at a cubic foot, weighing 62.5 lbs, the resulting product from the multiplication of the three quantities will give the pressure in pounds on the surface immersed. Let it be supposed, for simplicity, that water to the depth of 10 ft. has to be sustained by a vertical rectangular wall, as in Fig. 5. It is usual to take but 1 ft. length of the wall for the calculation, though it will not affect the result whether 1 ft. or 100 ft. be the length assumed. We then have the surface under pressure = 10 sq. ft., the depth of the center of gravity = 5 ft., and the weight of a cubic foot = 62.5 lbs., the product of which quantities gives us 3,125 lbs. pressure on 1 ft. length of the wall. But this pressure is not the whole of the force that the wall has to resist; the leverage that it exerts must also be taken into account. In the example under consideration—viz., that of a vertical plane with one of its sides coinciding with the surface of the water, as in Fig. 5—the whole of the pressure is so distributed as to be equal to a single force acting at a point one-third of the depth from the bottom. Thus, the total force to be resisted by the wall is 3,125 × 3.33

= 10,406, which is the moment tending to overturn the wall.

It is evident that a certain weight of the wall must be opposed to this overturning force; and as the height of the wall and the length are determined quantities, the thickness alone remains for adjustment. But as a rectangular wall in upsetting is considered to turn upon a single point, F, in the Figure—viz., the outer line of the wall—there will be a certain amount of leverage to assist the wall in resisting the pressure of the water. This leverage is the horizontal distance of the center of gravity of the wall from the turning point F, and when the structure is rectangular and vertical, it is equal to half the thickness. The amount of the wall's resistance will then be equal to the number of cubic feet in one foot of its length multiplied by the weight of a single cubic foot of masonry and by half the thickness of the wall. Taking $w =$ weight of a cubic foot of water $= 62.5$ lbs. $w' =$ weight of a cubic foot of brickwork, say 112 lbs., $x =$ thickness of the wall, and $h =$ the height, the conditions of simple stability will be fulfilled when

$$w^1 \times h \times x \times \frac{x}{2} = w \times h \times \frac{h}{2} \times \frac{h}{3} \dots (1)$$

$$\frac{w^1 h x^2}{2} = \frac{w h^3}{6};$$

and solving for x, we get

$$x = \sqrt{\frac{w h^2}{3 w^1}} \dots \dots (2)$$

the thickness of the wall $= 4$ ft. 4 in.*

A simple example has been selected for

---

* It will greatly facilitate all such calculations to make, once for all, certain reductions and simplifications: Thus, the weight of a cubic foot of water being 62.5 lbs., the thrust against a reservoir wall of height h, sustaining a body of water level with its top is, per running foot, $31.25h^2$, and the moment of the same is $10.42h^3$. Also the thickness x of a rectangular wall of density π per cubic foot, in exact equilibrium with this moment, is given by the equation, $x = \frac{4.565h}{\sqrt{\pi}}$. This shows that x varies inversely as the square root of the density. It will be found by trial that this formula gives the same result as (2), the only difference being, that in the one just given, all the roots have been extracted, except the root of the density of the masonry.

A good authority—Mr. Fanning—gives as a proper approximate thickness for a rectangular wall of density $= 140$ lbs. per cubic foot, $x = 0.55h$. It will be perceived that this is about 40 % additional beyond illustration; but of course a rectangular section of wall would not be found generally applicable in practice, nor would it be convenient to limit the dimensions of a retaining wall of whatever kind to the minimum that would sustain the pressure. If this principle of calculation be applied to ascertain the stability of a bank of earth with long slopes of $2\frac{1}{2}$ or 3 to 1, it can easily be shown that in

exact equilibrium. If we should make $x = 0.60\,h$, we would add 50%, and increase the moment of the wall by 19%, while the section would be increased only 9%, for the moments increase as the squares of the bases, and the sections, as the bases, only.

Such walls are not usually built with a rectangular section, but the best way to design a trapezoidal wall is first to determine the thickness of a rectangular wall with the desired coefficient of safety, and of the given height, and then transform it into one of trapezoidal section, having an equal moment of resistance. This may be readily done by using Vauban's rule, which is, that all walls with vertical backs and of equal resistance, have the same thickness at 1-9th of their height, to a very close approximation, and within extended limits. Thus, suppose we wished to transform a rectangular wall 27 feet high and 15 ft. thick, into an equivalent trapezoidal wall with vertical back, and top thickness of 6 feet. Join the outer extremity of the top with a point 3 feet above the base of the outer face, and prolong the line till it strikes the base prolonged. We shall then have a wall every case the resistance of the bank to overturning is greatly in excess of the horizontal leverage exercised by the water sustained.

The only theory, then, in any degree tenable, is that assuming a bank in yielding to the pressure of water to slide on its base. In order to conceive how this can apply, it is necessary to assume the embankment to be a rigid body resting, for a given length of its section, on a horizontal plane; and without any ad-

---

27 feet high, top thickness 6 feet, bottom thickness 16.125 feet, with a face batter of 4½ inches to the foot. The moment of resistance of this wall is slightly in excess of that of the assumed rectangular wall, while its section is less by more than 26 %.

Reservoir walls of considerable length—say over five times their height—should be reinforced by exterior counterforts, or buttresses, not further apart than the height of the wall, and carried up to at least half its height. The section of the wall should not be diminished on account of these buttresses, which are merely intended to give the long wall the same strength that a shorter one of the same section would possess. In all these calculations the adhesion of the mortar is neglected. Perhaps this omission is counterbalanced by the assumption that the wall is a monolith. If we wished to take account of the adhesion of the mortar, probably the best way would be to consider its intensity in pounds per square inch, as so much additional weight added to that of the wall.

hesion, or a very small fraction, existing between the surfaces pressed. The amount of the friction, however, is just the point upon which the whole matter hinges, and until it has been ascertained that the surfaces of earth that are carefully incorporated with one another have any such thing as a coefficient of friction, it is idle to pursue the investigation by a mathematical mode of reasoning. The conditions of stability will be satisfied when the horizontal component of the water's pressure against the bank will equal the weight of the bank, plus the vertical pressure exercised by the water to hold it down and multiplied by the coefficient of friction; but nothing is known of this coefficient, and consequently the equation remains incapable of solution. As a matter of fact, embankments do not slide bodily forward on their base when they fail, but give way from other causes than mathematical reasoning can supply. Landslips, it is true, to some extent support the principle that maintains the sliding of embankments; but, here, the circumstances are widely different. Landslips either take place when a mass of earth rests upon an inclined surface of rock, with an ample supply of water to lubricate the surfaces in contact, or else they are the result of cutting or embanking earth to a higher slope than the material will stand at; the infiltration of water also in this case is the chief agent in producing the effect, acting as a lubricant, and causing the earth to assume its natural slope. In each case the surface of separation is an inclined plane, an element that does not enter into the question of the stability of embankments, by either of the modes of reasoning above referred to. The principles that direct the design of embankments to retain water are not those that apply to the calculation of the forces to be resisted or the means to overcome them, any more than breakwaters and harbor walls can be designed on mathematical principles. The whole question naturally turns on what slope the material composing the bank will stand at.

If earth could be got to remain at a slope of 1 to 1, even though the embankment had no thickness whatever at top, it would be amply sufficient in weight to uphold the water in a reservoir. This, however, cannot be accomplished without the assistance of retaining walls, which would be found in most cases much more expensive than the additional earth required to increase the slope to the angle of stability; and therefore the section is so disposed that the earth shall stand both inside and outside the reservoir at such a slope as will be under all circumstances permanent. These slopes have been determined by long practice and by success and failure in pre-existing instances—that is to say, the limits have been laid down, for it is not to be assumed that all descriptions of earth will fall to exactly the same slope when exposed to the constant action of water or weather. Earth when subjected to the contact of water almost invariably loses a certain amount of its stability, and therefore it is usual to give the inner side of an embankment a longer slope than the outside. In most of the best existing examples the inside slope of the bank is either 3 to 1 or 2½ to 1, and it is rare to meet any departure from this rule. The outside slope may be designed at from 2 to 1 to 3 to 1, depending upon the character of the material, its power of withstanding the erosive action of the air, and the means used to protect the surface from being washed off or from crumbling away. In designing embankments, the impermeability of the earth is a matter that cannot be relied upon. There are, it is true, innumerable embankments now standing that have never allowed the escape of a drop of water from the reservoir, although no special precaution was taken to make them watertight. Of these India abounds with examples, the introduction of a puddle wall being in the older embankments of very exceptional occurrence. The earth was merely dug out close at hand, and carried by the work-people in baskets on their heads to where it was deposited, without any regard to the mode of disposing the material. The author has had occasion to construct a considerable length of levee, or embankment, on this simple plan for the protection of the country from the flooding of a river; and although, so far as he is aware, no flood has yet taken place to test the work, he has, from the study of existing examples, entire confidence in the result. The earth, so far as practicable, was disposed in layers, and before each was completed it was thoroughly consolidated by the tread of the workmen. It is not suggested that the puddle wall should be dispensed with in designing embankments, for the additional degree of safety, in most instances, will more than compensate for the extra expense it entails; but, in low embankments made of good retentive clay, the precaution of puddling is by no means a necessity.

In most of the best examples of embankments in England, the practice adopted has been to carry up the earthwork in layers of 2 or 3 ft. in thickness,

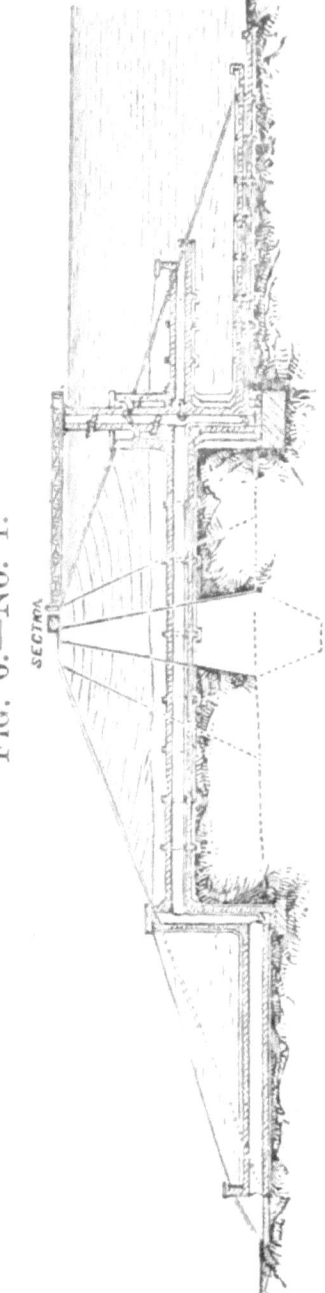

Fig. 6.—No. 1.

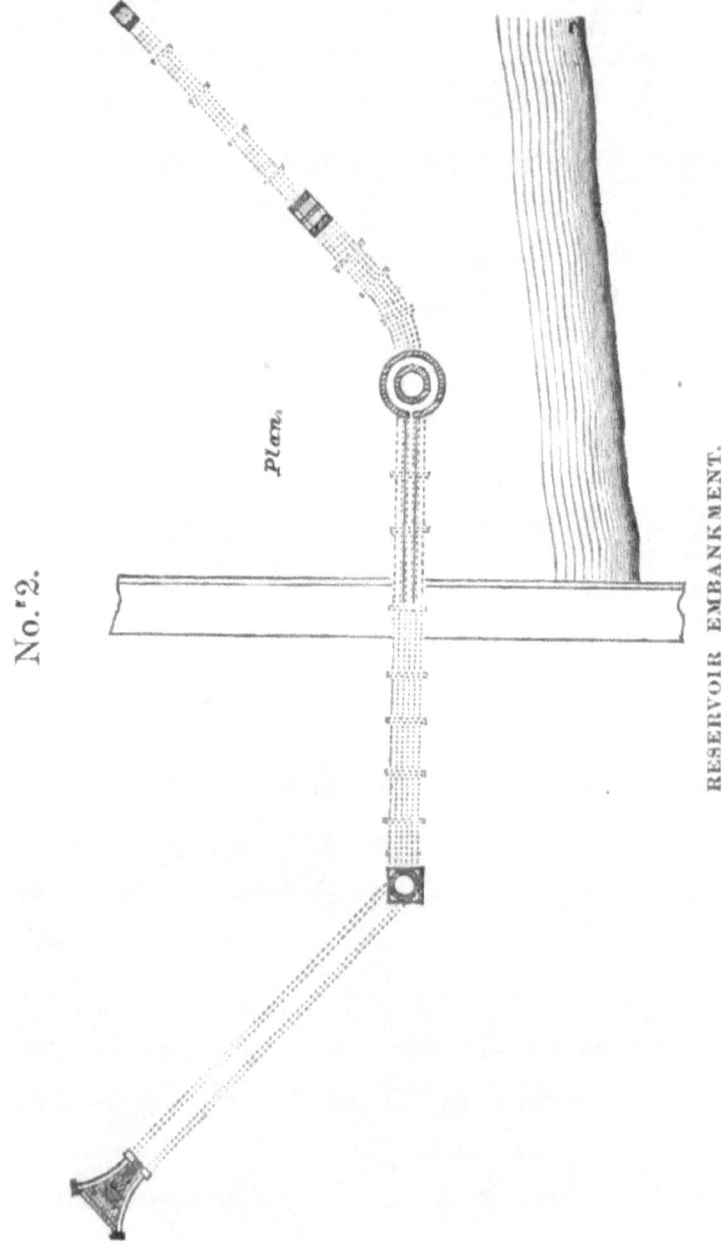

disposed in the manner shown in Figs. 6 and 7, and at the same time to construct in the center of the bank a wall of well-puddled clay, the foundation of which is carried down for whatever depth may be necessary in order to reach an impermeable bed of earth or rock. It is not in all situations possible to procure earth exactly suitable and in sufficient quantity for the construction of an embankment, and consequently it is usual and advisable to dispose the best part of the material—that is, the most retentive of water—in juxtaposition to the puddle wall, as indicated in Fig. 6.* In this example, the selected material is disposed equally at either side of the puddle; but, as its function is to withstand the admission of water, it would probably be more consistent, though less in accord-

---

* This practice is sound, but would in many cases be very difficult of execution, at least without considerable extra expense. When the borrow-pits are opened, the material is generally taken as it comes. The top soil, and all roots and sods should be first removed, and all stones larger than are allowed by the specifications are picked out and, generally, re-erved for the rip-rapping of the bank.

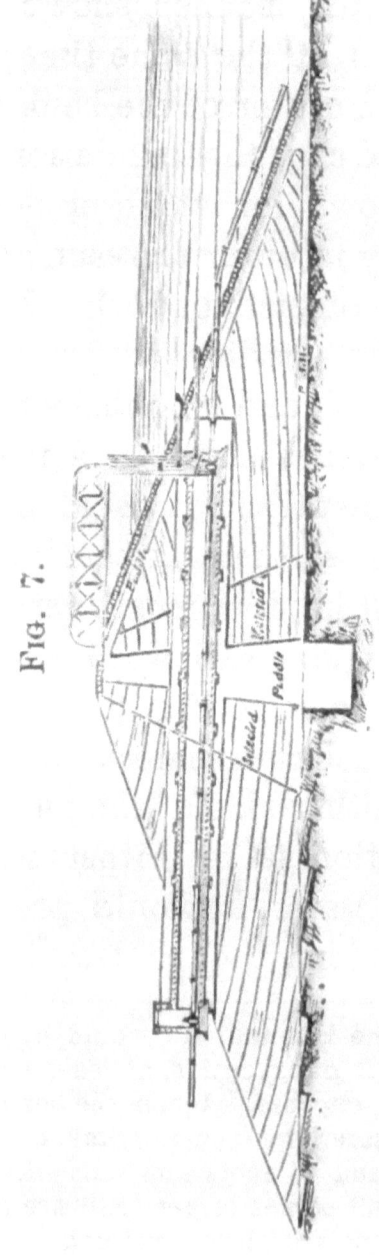

Fig. 7.

SECTION OF EMBANKMENT—BIDDEFORD WATER-WORKS.

ance with practice, to place all the selected material on the inner side. The practice of excavating the earth for an embankment from the inside of the reservoir is one that should not be followed without caution. Removing so large a mass of material would, no doubt, give a considerable increase of storage room; but sometimes the bed of a reservoir is covered by a layer of impervious clay that is of immense value, and if this be cut through or removed, it is quite possible that a bed of porous material may be met with sufficient to allow the escape of water when it comes to be admitted. In specifying for the dimensions of the puddle wall, a sound rule for adoption is, that it shall have a thickness of 10 ft. at the top water-line and increase in thickness to the surface of the ground at the rate of 1 in. on each side for every foot of height. Before any excavation is commenced, it will be essential to make a sufficient number of borings to ascertain the nature of the soil beneath the surface.

It may here be mentioned that professional men are not apparently agreed as to the principles to be kept in view in constructing reservoir embankments; and this want of concurrence never was more apparent than in the discussion that followed the destruction of the Dale Dyke reservoir, near Sheffield. Fig. 8 shows a plan of the embankment site after the catastrophe. The bank was 95 ft. high, with slopes of $2\frac{1}{2}$ to 1, and a top width of 12 ft. The puddle wall was 16 ft. in width at the ground-line, and tapered to 4 ft. at the top of the bank. This embankment, with the exception of the puddle wall, was composed of rubble stone and shale; an additional price having been given by the engineers to insure the use of the former material; which proves, at any rate, that this mode of construction was adopted on principle and not through ignorance or mistake. From the evidence given by the engineers of the company, it appears that it was, in their opinion, desirable that the inner part of the embankment should be per-

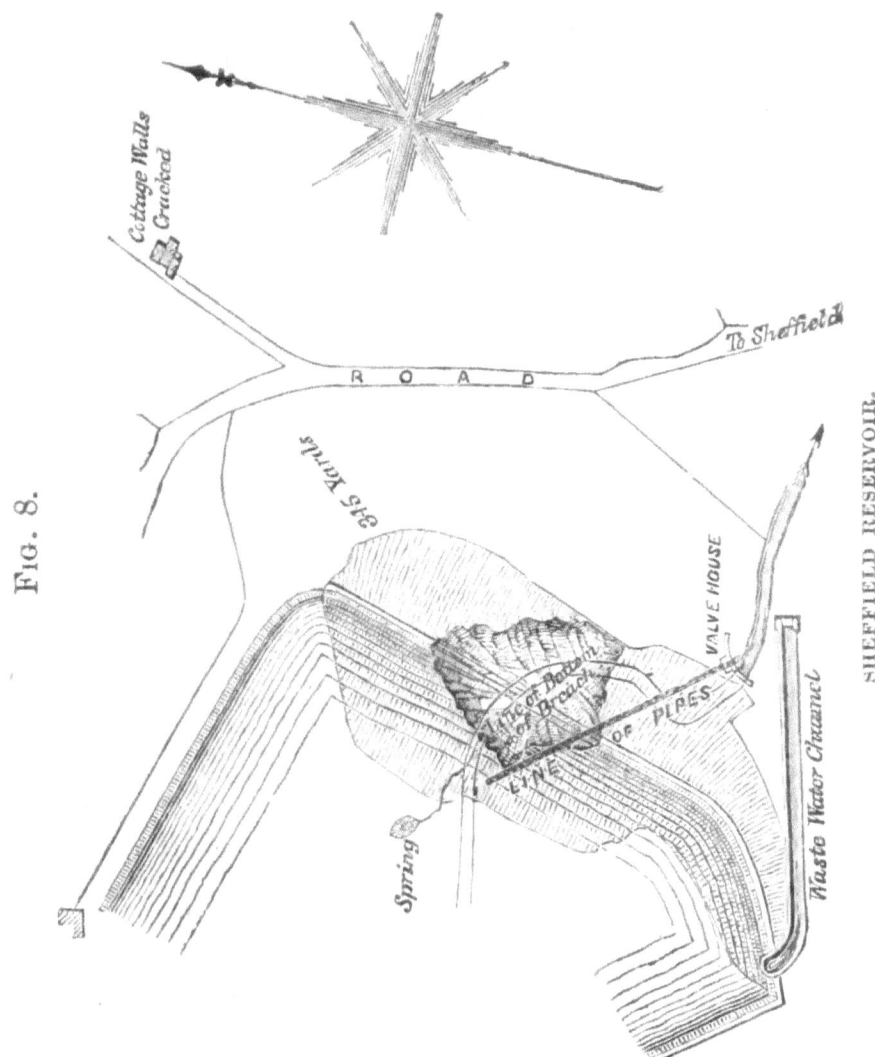

Fig. 8.

meable to water, because earth was much more likely to subside and slip than an open and less yielding material like stone. This mode of construction implies that the puddle shall be fully sufficient of itself to resist the passage of water, and that there is no necessity to relieve it of any part of the pressure against it. Of course, if a bank be composed of open work, every point in the face of the puddle is exposed to the full and direct hydrostatic pressure; and if at any point there is the smallest fissure or imperfection, the water has full power against it, and will, to a certainty, take advantage of such point to breach the dam. The assumption, then, of the constructors of this and the Agden reservoir evidently was that a puddle wall of some 25,000 sq. ft. of area was to be constructed without an imperfection of any kind, or a single weak point in the whole surface.

The obvious reason for employing puddle at all, in embankments, is to thoroughly close up any imperfection that may occur in the earthwork; it is in fact merely an accessory, and cannot be relied upon of itself to secure the embankment against destruction. If an embankment be constructed of good sound earthwork, properly executed, it is highly probable that the water may never penetrate half way through to the puddle wall, and probably, in the majority of examples, has not done so. Earthwork, however, is not always executed without imperfection; some decomposable material may be introduced, which, in course of time, dissolves, leaving a fissure; one part may be at first less consolidated than another, and, subsiding, lead to imperfection; or an embankment, be it ever so well constructed, may be burrowed through by moles, rats, and other vermin. It is to meet the first two of these sources of imperfection that puddle is used; and if, by such fissures as may occur in ordinary earth-work, water is admitted as far as the puddle wall, it can only exercise pressure against it at a few points, the puddle and earth being, in good work, so bonded and incorporated with each other that there is no space left for the water to occupy and press against the surface. Most who have read the account of the disaster that occurred in March, 1864, at Sheffield, will recollect how singularly conflicting the professional evidence on that occasion was. Some of the first engineers were ranged against each other in order to satisfy the public as to whether the failure of the embankment was attributable to bad engineering or to a landslip; and although the impression finally remained on the public mind that "there was not that engineering skill and attention to the construction of the works that their magnitude and importance demanded," the engineers were fairly divided in opinion as to the cause of the disaster. One section pronounced, without qualification, that the embankment gave way in consequence of a landslip, and entirely ignored the fact of the embankment being defectively constructed; whilst the other gentlemen gave their verdict dead against the company, and their mode of constructing water-tight banks.* The two diagrams, Nos. 6 and 7, may be taken as indicating the system of constructing embankments most generally approved of. The puddle, as will be observed, is carried up to the natural surface of the ground without any batter, and from that point slopes on each side to the top of the bank; on either side of the puddle is disposed, in concave layers, the most sound and retentive part of the material, and outside of all comes the ordinary earthwork.†

As a security against the eroding action of the water, and also against the inroads of vermin, the most desirable, as well as the most usual practice, is to pitch the whole of the inner surface of an embankment with stone, carefully laid by hand. Neglect of this precaution has led to the destruction of many embankments in other respects securely constructed, and even when ample height of bank above

---

* At the present day, it seems incredible that any other opinion should have been entertained.

† It may be doubted if disposing of the material in concave layers is good practice.

the surface of highest water was provided. In all ordinarily inclement weather the disturbance of the surface of a reservoir amounts to no more than a mere ripple; but when the surface is of large extent, and a severe storm blowing, the waves produced are such as to cause reasonable apprehension, and, in fact, have, before now, overtopped the bank and cut it down, till the water flowed over and caused the destruction of the work. In most cases, it will be necessary to leave about 5 ft. between the level of the highest water and the top of the embankment, and never less than 3 ft.

A mode of construction not very generally used, but apparently consistent with reason, is that shown in Fig. 7, the embankment for the Biddeford Waterworks. It consists in covering the whole of the inner face with a layer of puddle, with sometimes a layer of peat outside it. On some occasions it has been thought desirable to mix with the puddle a quantity of small stones or furnace cinders, by way of obstruction to vermin—a precaution that is by no means unnecessary. As an instance in point the author is reminded of a masonry dam in India that had to be pointed every year regularly, because the fresh-water crabs in the reservoir found it convenient and promotive of their development of shell to appropriate the mortar to their personal use. The joints were cleaned out as effectually at the end of each monsoon as if the work had been done to order.

The preparation of the foundation for an embankment is a matter requiring some care. The soil, consisting of grass, roots, etc., and other matters of a decomposable nature, should be carefully removed over the whole surface to be covered by the bank, and if any porous material, such as sand or gravel, be present, it must be removed, until a compact and water-tight bed is arrived at.* The bank

---

* This must be taken with some reserve. It is evident that carrying out this recommendation literally, would amount to digging out the whole site of the embankment to a depth equal to that of the center-wall foundation. All that can be done in most cases, is to remove the sods, roots, etc., from the site of the embankment, so that a fresh earth surface is laid bare, and commence the fill upon that.

must, in fact, be in contact with some sound and reliable material that will not admit the passage of water.

APPENDAGES OF RESERVOIRS.

Under this heading may be considered:

The whole apparatus for allowing the water to escape, including the pipes, the valve tower, and the culvert.

The waste sluices.

The waste weir or by-wash.

The most economical mode of discharging water from a reservoir is through a single pipe passing either through the embankment or immediately under it; but this plan cannot under any circumstances, be recommended, though it is some times found in existing examples. It is open to several grave objections, the principal of which, perhaps, is that the failure of a joint under the embankment from unequal pressure, or from whatever cause, will probably produce the destruction of the embankment, or at any rate, entail a serious interruption to the supply, by the reservoir having to be emptied, in order to repair the pipe. Buried in or under an embankment, a pipe is completely out of reach and out of view, and may be in a very defective state without its being possible to detect the imperfection.

In order to secure the satisfactory working of a reservoir as a source of constant supply, it is essential that the outlet pipes, valves, and all other appendages for controlling and regulating the escape of the water, should be accessible for inspection and repair. The usual mode of accomplishing this is to carry the pipes out through a culvert of brick or masonry of sufficient dimensions to admit a man. This culvert communicates with the valve tower, as shown in Figs. 6 and 7, so that there is a complete communication between the outside of the reservoir and the inside. When unavoidable, the culvert is carried straight under the embankment in the natural ground; but the safest and most generally approved mode of construction is to bring the culvert round the end of the embankment, where it will be out of reach of injury from unequal settlement; a source of no small apprehension when either culvert or pipes alone are carried under the bank. Where possible, it is an excellent plan to run a heading through the solid ground, lining it with brickwork and puddling it, forming a tunnel entirely independent of the embankment. The principal objection to carrying either the culvert or pipes through or under the bank is their liability to fracture from the enequal settlement of the earthwork. It would appear that their liability to damage cannot with certainty be insured by any reasonable depth of excavation, and is, therefore, generally disapproved of by the best authorities.

In the best constructions the culvert is situated half way or two-thirds up the embankment, and in such case the outlet pipes for drawing off the water in the reservoir act as syphons when the water surface has fallen below the culvert. Fig. 6 shows a plan, as well as a cross section, of a reservoir dam designed for general application by Mr. Rawlinson. Here the bottom of the culvert is about 25 ft. above where the inner slope of the embankment intersects the ground at the lowest point. The syphon pipe is also shown passing through the culvert; the horizontal culvert is connected with a shaft inside the embankment, in which are placed the valves for leading off the supply from the reservoir. The valves are made to be closed on the inside by valve spindles and screws, and the inlet pipes are closed on the outside by plugs which can be applied from the top of the valve tower. Thus the engineer has full command of the whole of the outlet works; all the pipes and valves are easily accessible and under perfect control, so that the supply can at any time be arrested for the repair of any derangement that may occur, even to the removal and replacement of all the pipes. The inlet pipes are shown in this example, as well as in Fig. 7, fixed at different heights in the valve-tower, the object of which is to draw the supply from the reservoir from points near the surface. The outlet pipe, passing through or under the embankment, may be connected on the inside of the reservoir by a flexible joint with another pipe of the same diameter, to the upper end of which is attached a float. This pipe is movable in a vertical plane, being controlled from lateral motion by the guide-posts. Such an arrangement admits of the water being drawn off from the surface, where it is least liable to be contaminated with impurities. Whatever arrangement be selected for drawing the supply off from a reservoir, the system of carrying the pipes, either with or without a culvert, through or under the embankment, cannot be sufficiently deprecated; they are, in such a position, beyond the reach of inspection, and, moreover, are very likely to induce leakage from the reservoir. It is usual to puddle carefully the culvert or pipes when carried under or through the bank, but, even with such a precaution, the water has under a considerable head a tendency to creep along the pipe, and, by soaking into the earthwork, may cause any one of the many evils that imperil and destroy embankments.

When embankments are not of great height, an exceedingly cheap and simple mode might be adopted for drawing off the water. This would be by laying a syphon over the embankment, as was done in the case of the middle-level drainage in Cambridgeshire, which syphon would at the inner side have a flexible connection with another tube having a float attached, as above described. Such an arrangement would apply in principle to heights not exceeding 30 ft., as the pressure of the atmosphere would maintain no greater height. In practice, however, the syphons cannot be worked with success at much above 20 ft., for it is found that after a short time, the flow becomes arrested by the collection of air in the upper part of the syphon, and it becomes necessary to pump the air out constantly, to prevent it from interfering with the flow, as it would do if not removed. It would appear a simple matter, where it is desirable to adopt a syphon, to utilize the power of the water flowing out for the purpose of getting rid of the air; it might easily be applied, through a small wheel and suitable gearing, to work an air-pump fixed at the highest point of the syphon, making the whole arrangement self-acting. The arrangement could be successfully applied to irrigation tanks in India, where the embankments are frequently less than 30 ft. Each leg of the syphon should be provided with a valve to retain the water, and when the supply was intermittent it would be essential to have an opening at the highest point of the syphon, and some appliance, perhaps an air-pump, for filling it with water in case of leakage.

To insure a constant discharge from a reservoir with a constantly varying head, several methods have been adopted; of these, one of the most ingenious is that used at the Gorbals Waterworks, near Glasgow. Fig. 9 represents a transverse section through the regulator-house, showing the arrangement by which the discharge is equalized. To the orifice of the outlet pipe, O, is fitted a square-hinged flap valve of wood, against which presses, by a friction roller, a lever, B, the arms of which are bent. To the upper arm is attached a chain that passes over a pulley, and is connected with a cast-iron cylinder or float, D, that stands in the reservoir, E, of slightly larger diameter. At the side of the entrance-door of the building is placed another cistern, G, of cast-iron, closed at top, and communicating by a pipe, R R, with the vertical pipe, H, which is in connection with the outlet pipe, and passes up the slope of the embankment, to carry away any air that may accumulate in the main. The cistern, G, is connected with the reservoir, E, by a pipe, K, which supplies water to float the cylinder, D. Now, it is evident that the discharge from the reservoir will be regulated by the position of the lever, B, and this again will be controlled by the height of the float, D. To regulate this height the supply from the cistern, G,

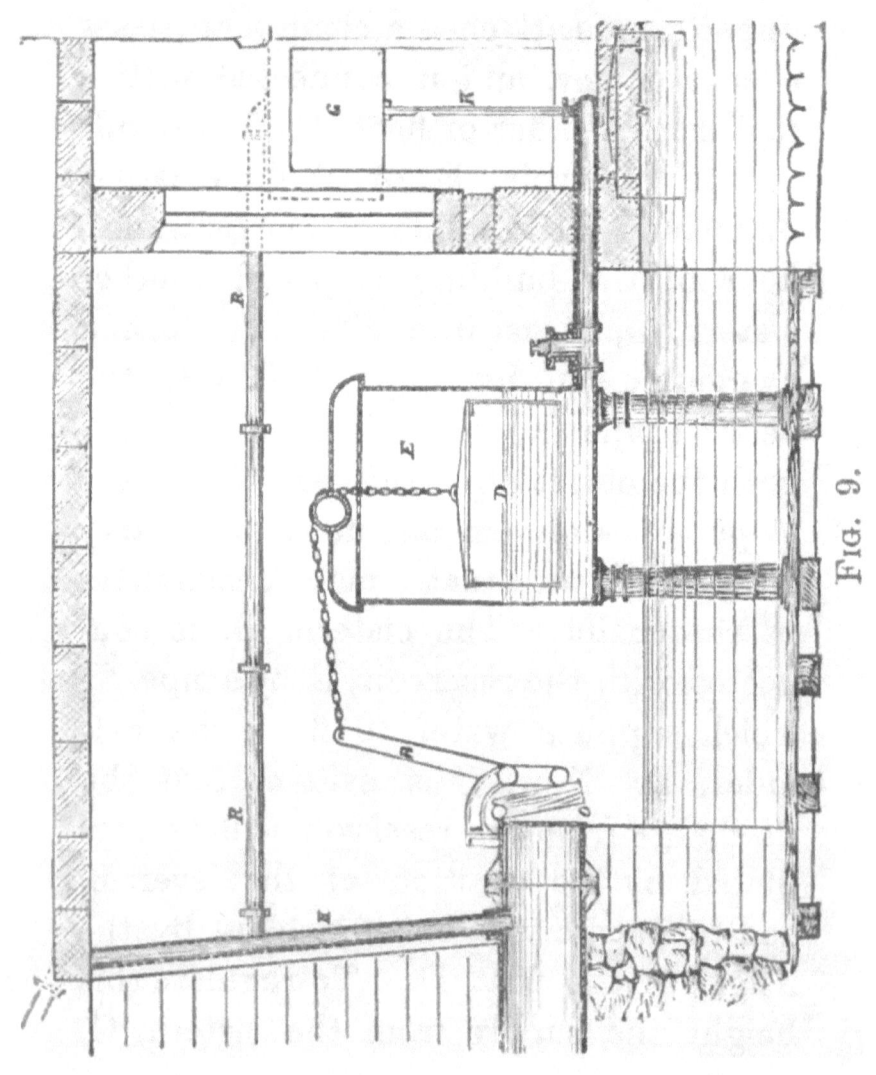

Fig. 9.

must be self-adjusting, or be regulated by the amount of water flowing away. The float, N, has attached to it a spindle, on which are fixed two double-beat valves that work in the vertical part of the pipe, K, one of which admits water from the cistern, G, into the cylinder, E, and the other allows the water to escape from the reservoir, E. Now, if the surface of the water upon which the float, N, rests should rise above the proper level, the float forces up the spindle, closing the supply valve from the cistern, and at the same time opening the lower valve. Thus the supply is cut off and the escape opened, enabling the float, D, to fall. The subsidence of the float closes more or less the flap valve, and checks the discharge, in consequence of which the surface of the water falls, and with it the float, N, which consequently opens the supply valve, and again admits water into the cistern, E. Thus an almost perfect equality between the consumption and the supply of water is preserved. It would appear that the same effect could be produced by connecting the lever directly with a float on the surface of the water, but such an arrangement would only apply when the pressure against the flap is trifling.*

It is essential that every reservoir should be provided with some means of getting rid of the excess of water that flows into it, and whether this provision be made by a waste weir, sluices, or waste pit, it is one that should not be omitted. The most advantageous position for a waste weir will generally be at some point remote from, and entirely unconnected with, the embankment, and occasionally a natural depression in the ground, as shown in Fig. 4, will afford remarkable facilities for the construction

---

* As a general rule, the discharge is regulated by opening the gates to a greater or less extent, according as more or less water is required. Indeed, it would seem unwise to have an automatic system, maintaining a constant discharge irrespective of a diminishing head. As the supply of water in the reservoir diminishes, it is better to let the draft from it diminish also, as a warning that the reserve is being drawn upon.

of an escape. The level of the crest of the waste weir with reference to the top of the dam will require to be carefully adjusted, the minimum difference of level being 3 feet, and the maximum about 10 ft., depending on varying circumstances. The height of the waste weir will, of course, regulate the top water level in the reservoir; and this must be fixed with regard to the probability of the embankment being over-topped by waves. The circumstances influencing the height of the waves in a reservoir are the extent of the water surface, the depth, and the amount of exposure to or shelter from wind, all of which will vary with each particular case. Under ordinary circumstances, the height of the top of the embankment above the crest of the waste weir should be for

an embankment 25 ft. deep, 4 ft.
" 50 ft. " 5 ft.
" 75 ft. " 6 ft.

and for greater height of embankment the difference of level may be proportionately increased. When the configuration of the ground does not afford any facilities for the construction of a waste weir after the manner described, sufficient provision for the escape of the overflow is made through a waste pit. This waste pit, or tower, is generally a circular structure built over the outlet culvert inside the reservoir, and serves equally for access to the valves and for the escape of the flood water. With regard to the capacity of the waste weir or waste pit, whichever be adopted, it will be necessary to make ample provision for the discharge of the sudden accessions of flood water that reservoirs are subject to, and which so seriously imperil their safety. To provide for this there is an empirical rule amongst engineers that is supposed to suffice for the most urgent contingencies. It states that there shall not be less than 3 ft. of length of overfall for every hundred acres of gathering ground, but it is obvious that to proportion the length of the waste weir to a given area of country in all cases would be unreasonable.*

The discharge over the weir will not depend only upon the quantity of rain falling on a certain area of ground, but also on the extent of the reservoir as compared to the gathering ground, and on the flat or precipitous character of the basin. The only safe mode, then, of proportioning the length of the escape will be to ascertain with exactness what the discharge of the stream or streams flowing out of the reservoir was during the greatest known flood, and then fixing upon an arbitrary depth for the water to flow over the weir, say 2 ft. or 3 ft., to calculate what length of overfall will suffice for the discharge of the excess water. In India, where large waste-weir accommodation is essentially neces-

---

* The above rule might answer up to say 3 square miles of drainage area, or gathering ground. Beyond this it would give excessive lengths. For large areas, up to say 50 square miles, 3 ft. per square mile would be nearer the mark. A rule that would fit fairly well, all cases, would be: Length in feet = 20 times the square root of the number of square miles of drainage area.

sary, while it is equally a necessity to save every gallon of water that is possible, it is a common practice to form a temporary dam, of earth and sods, on the top of the waste weir; this serves to pond up some 3 ft. or 4 ft. of water over the whole surface of the reservoir, and does not imperil the security of the works. In times of heavy floods the water rises and overtops the temporary dam, and no sooner does so, than the whole is carried away, and the water in the reservoir quickly subsides.

In works designed for the supply of towns, it is sometimes necessary to make provision to arrest the entrance of floodwater into the reservoir, as the streams may come down charged with large quantities of matter in suspension that would injure the purity of the water for domestic consumption. These streams may be diverted and carried round the margin of the tank past the dam, and can be admitted into the channel of the stream, or be utilized for mill power. On the Manchester Waterworks, are constructed across the mountain streams, weirs of an ingenious design, for the purpose of separating the flood-waters from the ordinary flow. The dimensions are adjusted from observations of each particular stream, so that the discharge up to a certain amount will take place into the channel for the supply of the town; but when the discharge increases, and the water becomes turbid, it has sufficient velocity to carry it over the opening, and flows down to the compensation reservoir for the supply of mill power.

In determining the dimensions of a weir of this kind it is first to be ascertained what the mean velocity of the water flowing over will be for a given depth of water, h, above the crest. The mean velocity, v, will be

$$v = \frac{2}{3} \times 8.024\sqrt{h} = 5.35\sqrt{h}.$$

If the vertical height of the crest of the weir above the point to be overleaped by the cascade be called x, the distance across will be

$$y = \frac{2\,v\sqrt{x}}{\sqrt{2\,g}} = \frac{4}{3}\sqrt{x\,h}.$$

Before concluding, it will be well to give a brief consideration to the causes tending to the failure of embankments. The foregoing remarks will, in suggesting the best mode of construction, have anticipated much that might be said on the subject of failures; but there are a few points, the recapitulation of which the importance of the subject demands.

There are unfortunately on record, accidents, if they can be so called, from the bursting of embankments, that, if estimated by the loss of life attending them, are as appalling as anything within the memory of man. Thousands of human lives have been sacrificed to ignorance and false economy, as well as in some instances to natural defects that it would have been difficult to foresee.

The existence of springs on the site of an embankment is an undoubted cause for apprehension, and considerable care should be taken to carry all water from this source away;* that it may not, as it certainly will if not checked, force its way between the surface of the ground and the seat of the embankment. In doing so there is every probability that the earth of the embankment will be washed out by constant trickling, till a fissure is formed of sufficient dimensions to render the destruction of the bank a certainty, if the water from the reservoir should ever penetrate so far. As a provision against this source of injury, all springs found on the site of an embankment should be taken up and carried away in proper drains sufficiently and securely puddled. Thus the water is confined to a single channel, and has no tendency to soak into the earthwork and blow it up in endeavoring to escape. In embankments of all kinds the presence of water is a most serious evil, and

---

* Where to? All these springs, and all percolations whatever, must be cut off and prevented from traversing the embankment by a water-tight puddle or center wall, which is the only safeguard against this danger, to meet which is one of the principal objects of such wall.

one by which may be accounted for, some of the most extensive land slips that are on record. It is erroneous to assume that when water is the active element in producing disruption in an embankment or mass of earth of any kind, that it only acts as a lubricant between the surfaces in contact. The truth is, the bulk of earth is sensibly affected by the amount of moisture in it, as is seen in the subsidence of newly-formed railway banks when exposed to rain. If, then, a sufficient quantity of water find its way into the center of a bank that has been put together in a comparatively dry state, it will rise and soak into the earth until at length what was a solid mass becomes semi-fluid, settles into a smaller space than it before occupied, and, as a consequence, will leave a vacuity above it. The inevitable result is the subsidence of the superincumbent earth; but instead of resting, as at first, on a resisting material, it floats, so to speak, on the semi-fluid mass underneath, and having little or no friction to overcome, slips away to a lower angle than it before stood at. Natural springs, therefore, whenever they occur, must be dealt with carefully and completely. Exactly similar effects to those produced by natural springs may result from the defective practice of carrying outlet pipes through or immediately under embankments. Be the pipes ever so well puddled, there will be a tendency to trickling along the line of their direction, and assuredly if this trickle makes its way to the center of the bank it will carry mischief with it. It is true that springs are occasionally found issuing from the foot of an embankment, without after several years causing any appearances to justify apprehension. The Doe-park reservoir is an example in point, and though at one time fears for its safety were entertained, the embankment is still standing, and, so far as the author is aware, the spring is still trickling away. An engineer of eminence was called upon to report upon the state of the works, and gave his opinion that, as the spring came away without any earth in suspension, there was no mischief taking place, and that the work was in a safe condition. There is no doubt that embankments in this condition require to be narrowly watched, although the presumption may be that, having lasted for several years, they will continue in safety.

The empirical and unscientific mode of proportioning the length of waste weirs has proved before now a source of danger and destruction to embankments, from the space afforded not being sufficient to discharge the excess water without the surface rising to such a height as to top the embankment. To avoid risk, the stream must be gauged with great care, and the discharge calculated for the greatest known flood; and if with a given head the length of the weir be adjusted to discharge this amount, or a little in excess, there will be no risk to the embankment.

Regarding finally the whole subject, the danger that may result from careless or unscientific construction, the large outlay entailed in the establishment of storage works, and the benefit that may accrue from them whatever their purpose may be, the subject cannot be undertaken on merely rational grounds. Its successful application will rest alone on the study of the question in its scientific details, and an ample practical experience.

DISCUSSION.

Mr. H. P. Stephenson said he entirely agreed with the author as to the impropriety of carrying a pipe through the embankment of a reservoir. He would extend his objection to the passing of a culvert through the embankment. If the culvert were laid on the natural ground, they would avoid the risks pointed out by the author, either of the settlement from the joints of the pipe, or of the water creeping along between the material and the pipe. He believed that the true principle of construction for reservoirs was the placing of a good puddle dam in the center, and on each side of this dam layers of earth well punned in. One reason why he should prefer the puddle wall in the center was that there was less tendency in the puddle to slip in such a position than when laid on the slope.

Mr. Albert Latham agreed with Mr. Stephenson in his remarks as to the pipes and culverts; but he thought it was an open question whether the puddle wall should be in the center of the dam. He had a strong opinion that it should be on the face of the dam.

Mr. Cargill said that he believed that the reason the puddle wall was not required in Indian embankments, referred to by the author of the paper, was that the earth seemed to have been thoroughly consolidated by the continual trample of people upon it. That thorough-consolidation was the great point in all puddling, and it was on that account that specifications were generally so stringent as to the thickness of the layers of the puddle. As to the position of the puddle wall, he could not see the particular value of having it in the middle of the dam, and he thought that a far better place for it would be the face, because the object of the puddle wall was to prevent the infiltration or the escape of the water.* This could be effected by puddling the whole slope right down to the permanent strata. The puddle wall was not required to promote the stability of the dam. The question of putting pipes or culverts under the dam required more consideration. It was alleged that the putting of a naked pipe through the dam of the Bradfield reservoir was one of the causes of its bursting. In some very large waterworks now being constructed in Dublin there were two distinct sets of main pipes, and they were laid in two large culverts at the bottom of the dam. The culverts were large enough for a man to walk upright in them. If the foundation were well looked after, there

---

* As the center wall, of puddle or masonry, is the *citadel* of the dam, it would seem clear that it should be protected by being placed within the embankment. Placing the puddle on the outside, in the shape of a face covering, would seem to invite such a disaster as that mentioned as having occurred at Sheffield.

would be no fear of the arch or dome of the culvert giving way in consequence of any inequality of pressure above it, as, if properly constructed, an arch would stand any amount of pressure short of what would crush the material.

Mr. Baldwin Latham said he could not agree with Mr. Jacob that a dam could not be constructed from theoretical deductions; for unless regard was paid to theoretical considerations there might result either a deficiency of strength or a waste of material and labor. In the dam shown in the drawings, and designed by himself, the pipe did not run through, but on the outside of the dam, on the solid ground. It was a well received opinion among engineers that if you had a pipe or culvert running through an embankment, that pipe or culvert would be unsafe. He believed that well made and properly tested pipes were quite as safe as culverts when in the solid ground. A pipe was simply a small culvert made of iron instead of brickwork. In cases in which there was a tendency for the water to creep along the outside of the pipe, that might be stopped by having projecting flanges on the pipe. The same creeping of water might take place along a culvert as along a pipe. With regard to the slope of a dam, the inside slope should be greater than the outside slope, because the greater would be the stability of the dam, and the water would have less destructive effect on the dam; he had effectually prevented leakage by the use of socket-pipes. The square projection of the sockets was always presented to the reservoir, and the pipes were laid in the virgin ground. It was very bad practice to lay the pipes in made ground, and especially through a dam. Pipes laid under a dam should be tested under pressure after being laid and before being covered up, so that any defective joint might be discovered. In cases in which he had laid pipes through dams, they had been so tested, which resulted in good and effective work; but he was bound to say that, if the pipes had not been tested *in situ* the result would not have been satisfactory.

Mr. Schönheyder said that Mr. Jacob had said that wherever springs occurred they should be well carried away. He (Mr. Schönheyder) wished to know how a spring was to be prevented from diffusing through the earth.

Mr. Hendry said that he had seen pipes which were laid through embankments, but had never seen one that was perfectly tight. It was almost impracticable to make it so, owing to the continuity of the puddle being disturbed at the point where the pipe passes through.

The chairman asked what was the largest diameter of pipe Mr. Hendry had seen used.

Mr. Hendry replied that the largest was 18 in. He had heard of several methods being tried, but he did not think it was possible to prevent leaking, more or less, from the reservoir along the outside of the pipe. He should like to be informed how it was possible to connect the puddle with the pipe; if the pipes be laid in the natural ground below the foundation of the embankment, then there is no fear of leakage, provided the pipes are properly laid.

Mr. Jacob, in replying to the discussion, said, that in the opinions that had been expressed there were but few points of disagreement with those that he himself held. He could not agree with Mr. Latham in his belief that embankments could be calculated on mathematical principles. In order to deal with embankments theoretically, they must be regarded as rigid masses, and be assumed to rest upon a horizontal plane. It could be shown mathematically that a rigid body of the same specific gravity as ordinary earth need not present the same section as is usually given to embankments, in order adequately to resist the pressure of water. A right-angle prism with the hypothenuse resting upon the plane would be quite sufficient to resist the pressure of water, even supposing the surface of the water to coincide with the upper edge of the prism. The reason of giving long slopes to an embankment is discoverable from the fact that banks, when exposed to the action of water, are found to waste and slip away to such an angle as will withstand the action of the water. The chief reason of the failure of embankments is the infiltration or soaking of the water from the inner side, which renders the material semi-fluid and causes it to subside into a smaller space than it originally occupied. The superincumbent mass then sinks and allows the water to overtop the embankment. The earth used for making embankments in the Deccan and in parts of the Madras Presidency in India is of a most suitable quality for the purpose. It is what is called "black soil," being very dark in color, and of a highly argillaceous character. The color is, no doubt, due to the presence of carbon. The clay makes most excellent puddle; but, no doubt, the consolidation produced by the tread of the work-people is the real secret of the earth resisting the pressure of water so successfully as it does. In North America, the levees for protecting the country from flooding by the Mississippi are sometimes constructed simply of sand; and are found, for the most part, sufficient for their purpose. As regards carrying away springs from the seat of an embankment, there is no difficulty in ascertaining where they exist when the ground is laid bare, as they are generally well-defined streams. Before the earthwork is commenced it is necessary to construct drains of masonry, or brickwork, or to lay iron piping to carry away the water clear of the work.*

The chairman said that the paper of Mr. Jacob was a very interesting one, and the subject was one which, during the last year or two, or, he might say, within the last week or two, had commanded the attention of the whole body of engineers. Last session a special Act of Parliament was passed, that all reservoirs and embankments should be constructed

---

* We are again compelled to ask: Where to?

to the approval of the Board of Trade. The subject of irrigation in India, which was alluded to in the paper, was one of vital importance. There was no question that the only means we had of irrigating that country in an efficient manner was by the construction of reservoirs.

Additional Remarks by the American Editor.

The importance of storage reservoirs for the purpose of equalizing the flow of water furnished by streams, is so generally recognized that they must be considered as forming essential parts of all well-planned systems of water supply, excepting those maintained by bodies of water of such large relative dimensions that their minimum flow in the driest seasons exceeds the maximum consumption.

The admirable paper of Mr. Jacob, which forms the first part of the present volume, covers the ground embraced by its title, and, together with the subsequent discussion, forms a body of information of the highest interest and value. The art of hydraulic engineering, however, is a progressive one, and at the present day there is much which may be added to the original paper, particularly from the point of view of American engineering, which it may be fairly claimed is likely to enhance its usefulness.

In planning storage reservoirs, there are two questions which naturally command attention from the very start: How much stored water does our supply require? And how much can our available resources be counted upon to furnish? In answer to the first of these questions, I cannot probably do better than quote the following statement made by Mr. Pole (Proceedings of the Institution of Civil Engineers, 1884-5) as regards England: "The general judgment of experienced practitioners appears to be, that for large rainfalls, a storage of 150 days' supply, or even less, will suffice; but in drier districts it may be necessary to go as high as 200 days." This rule will, I think, hold good for this country also, and we may, as a general thing, consider that a water supply which comprehends a storage equivalent to 150 days' consumption, is in a good position for the maintenance of a constant supply at all seasons in districts enjoying an average rainfall. It will be understood that this amount of storage applies to cases where it is proposed to use the whole, or greater part, of the total yield of the stream.

As to the second question, the answer depends upon the minimum annual rainfall over the basin drained by the given stream, and the amount of this rainfall—deduction made for losses by evaporation, percolation, etc., as well as for consumption, which is available for storage. In regard to this matter, Mr. Jacob dwells upon the necessity of ascertaining the amount of annual rainfall, and also of carefully gauging the stream, in order to establish its discharge at different seasons of the year. Now, it is evident that these operations, to be of any real value, require a long period of time for their accomplishment, and can be therefore, in the great majority of cases, of only partial utility. Fortunately, we can frequently dispense with any consideration of this point, for our stream may be of such size, and may drain so large an area, that we need not be at all troubled as to its ability to fill our proposed reservoir to the desired capacity. On the other hand, however, cases often present themselves when it becomes a question, if we build a reservoir to contain the required amount of water, whether we can be sure of filling it every year. It is not safe, as a general rule, and in average locations in this country, to count on more than 12 inches of available annual rainfall, for storage purposes, and even this limit, if great interests are at stake, should be approached with caution. Twelve inches of rainfall will furnish 27,878,400 cubic feet, or a little more than 208.5 million U. S. gallons per square mile of drainage area, and, for a round number, 200 million gallons per square mile may be considered as the maximum that it would be safe to *count on*, although it is very probable that there would be times every year when water ran to waste over the spill-way, indicating that the capacity of the reservoir was not as great as the flow of the stream would warrant. It might therefore be wise, in cases where it was important to store every available gallon, and worth while to spend a good deal of money to make sure of doing so, to increase the relative capacity of the reservoir.

In the Croton basin the average yearly precipitation is almost exactly 46 inches. The very careful observations made by the Croton Aqueduct Department, and extending over a long period of years, indicate that in this basin each square mile of drainage area will furnish an average water supply of at least one million gallons per day. In districts of similar geological character, and with equal average yearly rainfalls, the same rule may be considered to hold good. Under such circumstances, a reservoir capacity of 200 million gallons per square mile would furnish storage for 200 days' supply.

It will be seen that the three points governing the question of storage capacity are—the drainage area, the available rainfall, and the daily supply which it is desirable to maintain.

These preliminary questions being settled, the next point coming up for decision will be, the best location for the dam which is to form the reservoir. The approximate location can generally be readily selected: The natural features of the ground usually clearly indicate the point near which the dam should be built. In order to determine the exact position, the whole probable area should be crosssectioned in 20 feet squares (with additional points where needed). This work will not only show the precise line where the longitudinal section of the dam has the smallest area, which, other things being equal, will be the best line, but will also preserve a record of the original surface of the ground, from which the amount of work done at any time can be readily estimated.

The height and location of the dam being thus settled, the next important question is, What sort of a dam shall we build? It will be observed that in Mr. Jacob's paper, he confines himself to the consideration of earthen dams only. These dams can usually be built cheaper than masonry dams, and have this advantage, that they admit of being built in places where masonry dams would be unsafe. It will very seldom be found advisable to build a high masonry dam on anything but a foundation of solid rock. When such natural foundation is not found, and it is still determined to build a masonry dam, the only resource in order to secure safety, is to carry the foundations down to a depth which will very greatly increase the cost of the work.

The dams spoken of by Mr. Jacob are of a type very prevalent in England and India: namely, earthen dams with a puddle core. Such dams are rare in the United States, where the puddle core is commonly replaced by a masonry center wall. The object in either case is, primarily, to establish an impervious cut-off against any possible percolation through or under the earthern embankment. I think there can be no doubt as to the superior efficiency of the masonry wall, and I doubt if there is as great an economy in the puddle, as would at first appear probable. In the first place, proper material for good puddle is not always obtainable, and even when it is found in abundance near by, its proper preparation and placing are matters requiring a good deal of careful and expensive manipulation. Good puddle should resemble, when made and placed, in character and composition, an unburnt brick. When we read, as we do in Mr. Jacob's paper, and elsewhere, of the great precautions necessary in putting in a puddle wall, and the disastrous consequences attendant upon some apparently trifling neglect in doing so, I think most engineers intrusted with the designing of so important a work as a large storage reservoir, would hesitate before risking these consequences in order to effect an economy which perhaps might eventually prove to be not so great as was anticipated.

There is another particular in which the masonry center wall presents a great superiority to the puddle core. One of the weakest points of an earthen dam lies in the neighborhood of the conduit used for drawing off the water. A perusal of Mr. Jacob's paper and the discussion shows what importance is attached to having this conduit form a perfectly water-tight connection between the inside and outside of the reservoir. It is clear that the existence of solid and water-tight masonry center wall, running through the entire length of the dam, and firmly and deeply imbedded in the banks of the valley on each side, affords an excellent opportunity for making all the necessary connections in a satisfactory manner. The masonry culvert through which the water passes, and the gallery containing the pipes, may be bonded in with the center wall and made to form a part of it in such a way that the possibility of water following along the outside of these structures is wholly precluded, and one chief danger of destruction of the dam entirely averted.

Having now decided upon the character of our proposed dam; viz:—an earthen dam with masonry center wall, or cut off, it is next in order to consider its design, dimensions and accessories.

As regards the latter, the spill-way merits the first mention. The dimensions of this portion of the dam must be ample, and sufficient to safely pass all the water which may come to it in times of heaviest freshet, without the possibility of its overtopping the dam. As to the proper length of the spill-way, it is perhaps impossible to lay down any fixed rule, or to attempt to make it a given function of the water shed.* If, in designing a dam, we find any existing dams in the neighborhood, upon the same stream, or, failing this best indication, any railroad bridges through which the whole of the stream has to pass, we may have a good opportunity of ascertaining the amount of water passing off in freshets, and proportion our spill-way accordingly. If none of these indications are to be

* See page 87.

found, the next best course to pursue (and this should be done in any event, as a check), is to ascertain the dimensions of the spill-way of some existing dam, built elsewhere, but upon a stream of about the same drainage area as the one in question, and, if possible, of the same character.

When a natural spill-way cannot be found, such as Mr. Jacob speaks of, and which, of course, is always preferable, its construction of stone masonry is always a very expensive piece of work, and the tendency is therefore to reduce the length, and provide for a deep wave passing over it. This kind of economy should not be pushed too far; it is better and safer to provide a long spill-way over which a comparatively thin sheet of water shall pass.

The dimensions of this important part of our dam having been settled, the next point to consider is what relative position it should occupy. If the sides of the valley are of rock, even if not of the very soundest and most solid character, there will be an economy in placing the spill-way at one end of the dam, as its height will thereby be diminished. If, however, the sides are of earth, the safest place for it is directly in line with the stream, as it would be dangerous to discharge a large volume of water upon the unprotected hillside, and any attempt to protect it against wash, by means of walls and paving, will generally involve an expense equal, or nearly so, to the higher structure. This is a point, however, which is by no means to be taken for granted, and in cases where great economy of construction is necessary, the ground should be carefully examined, and comparative estimates made.

As regards the form of the spill-way, the curved form adopted for the old Croton dam, and the dam on the Bronx, at Kensico, N. Y., is no doubt the best, but its great cost in the way of cut stone voussoirs will generally preclude its use except for municipal work, and where economy is studied, the form adopted will generally be that of steps, or offsets.

A good form of spill-way is shown in Fig. 72, page 382, of Fanning's "Water Supply Engineering." A very good and massive form, which I have had occasion to adopt with some modifications for a spill-way about 30 feet high, is described as follows: Back, vertical; top width, 7 feet; batter on face, an inch and a half to the foot, for 8 feet, making a thickness of 8 feet at the bottom. Thence, steps ranging in height from 15 to 22 inches, and following a general slope of 45 degrees. In this way, it will be perceived that at any given elevation, the thickness of the wall is always equal to its height above such elevation.

The waste pits and sluices spoken of by Mr. Jacob as substitutes for a waste weir, are not, I think, to be recommended. Nothing can be better than a straight opening and clear escape for the surplus water.

As regards the means adopted for drawing off the water, they should be as simple as possible, with good facilities for inspection and repair. Those mentioned by Mr. Jacob seem to be unnecessarily complicated. My own opinion is strongly in favor of having a cast-iron pipe or pipes built into the center wall, and extending outside of the dam, in an arched gallery, founded in the natural formation sufficiently large to admit of free circulation all about the pipes. Within this gallery, or gate chamber, are the gates, two upon each pipe, the inner one, or one nearest the water, to be kept habitually open, and the delivery of water regulated by means of the outer one. If any accident occurs to this gate, the inner one is closed, and it can then be got at for repairs. On the inside, a tower is built, one side of which is formed by the center wall through which the pipes run. This tower, or well, is rectangular in shape, and its sides are from one foot to eighteen inches outside of the pipes; that is, its inside width is from 2 to 3 feet greater than the outside diameter of the reducer of the largest pipe, so as to afford a chance to lead and calk the reducer when set in the cut ring stones which surround it. The length of the tower may be from 10 to 15 feet. It contains two sets of cut stone grooves, in which stop-plank can be placed. By this means, should it ever become desirable to get at the mouth of the pipes or reducers without first emptying the reservoir, the stop-plank can be placed, and, if necessary, the space between them puddled, and a water-tight coffer dam is thus made, inside of which work can be carried on while the reservoir is full of water. From this tower, a masonry culvert, also provided with stop-plank tower, if thought necessary, or an open passage, with wing walls, extends through the embankment. The gate chamber, tower, wing walls and spill-way should be, if possible, grouped together, with a view to economy of materials and increased strength, and an effort should be made to so locate the work that these important features should be set upon the most favorable natural foundation.

As regards the center wall of masonry, it is, of course, impossible to lay down hard and fast rules for its proportions, any more than for the other parts of the dam. But certain general principles may be established, to do which it is first necessary to get a clear idea of the precise function of the center wall, and of the part it is to play in the dam. As we have seen, it is intended, primarily, to afford a water-tight cut-off, to arrest any percolations which may reach it, by trickling through the bank. Indeed, we may consider the center wall as constituting the dam proper, for it is to the center wall that we finally look for the retention of the water within the reservoir. Regarded in this way, the earthen embankments on each side are only provisions for keeping the center wall from being thrown down. In point of fact, however, we do expect more than this from the embankments. We expect the inner embankment to be very nearly impervious, and of itself to be almost, if not quite, sufficient for the retention of the water. In any event, we expect it to be a powerful auxiliary to the center wall, by keeping the deepest water well back from it, and increasing the distance that a given drop of water would be obliged to travel in order to pass underneath the wall. In this way, the embankment is equivalent to a deepening of the foundation of the wall. Moreover, water reaching the center wall by traversing the embankment, comes to it in the form of a percolation, modified by capillary action. The object of the exterior bank is mainly to keep the wall from being thrown out, but it serves an excellent purpose also in smothering down any slight percolations issuing from underneath the wall, and still further increases the distance that a given drop of water must travel in order to pass freely from the inside to the outside of the reservoir.

As to the proper height to give to the center wall, although it is highly probable that in many cases it is not necessary to carry it up to the full height of the surface of the water, when the reservoir is full, we cannot say that an earthen dam is absolutely safe unless the wall is carried at least as high as the lip of the spillway. It is still better, in very high dams, to raise it as high as the extreme elevation of flood freshets. As to its thickness, economy will prompt a reduction to the narrowest limits, but it must not be forgotten that, although it depends on the banks for support, rather than to its own moment of inertia, yet, owing to the fact that some movement may take place in the banks themselves, it is liable to be subjected to unbalanced pressures, and should therefore be, to some extent, self-supporting. If we start with a top thickness of 5 feet, and increase the thickness, by off-sets, to the extent of 2 feet additional, every 10 feet as we go down, we shall have dimensions which, in a great majority of cases, will be abundant to afford the required strength.

A still more important question in regard to the center wall, is the depth to which its foundations should be carried. For this there is absolutely no rule, and the question must be decided according to the engineer's best judgment for each particular case. Of course, if a rock foundation is encountered, the problem is greatly simplified, for we have only to remove the loose and disintegrated surface, until we come to clean, live rock, and place our bottom course upon that. The uncertainty occurs when our test pits reveal the existence of coarse and permeable strata, when our only resource is to carry our wall well down. In some cases such walls have been put down to such a depth that in places the distance below ground is greater than the height above, and frequently sheet piling has been resorted to. This, however, it is best to avoid, if possible.

In general, it may be said that the finer the material, the better it is adapted to serve as a footing for the center wall. Clay, fine gravel and sand, and quicksand when found at a sufficient depth to prevent its being forced up at the sides, are all excellent materials to build such a wall upon. Where a compact, watertight material is found, it is needless to carry the footing courses to any great depth below the surface of the ground, but when loose cobbles, coarse gravel and shingle, or any other ground affording an easy passage for water, is encountered, the only safety is in depth of foundation.

It now remains to say a few words respecting the earthen embankments. On page 85 Mr. Jacob gives a little table of the proper heights for the top of embankments above the crest of the spillway, or waste weir. It is very important that this height should be amply sufficient to prevent the embankment being overtopped during a freshet. Of course, the length of spill-way is a factor in this matter, but the top of the embankment should be kept well up, and I should be disposed to add a foot, or even two, to the figures given by Mr. Jacob, making them respectively 5, 7 and 8 feet, and I would change the maximum from 10 to 12 or 14 feet. The top width may vary, according to the height of the dam, from 5 to 20 or more feet. Perhaps a fair approximate rule would be, to give the top of the embankment such a width that, taken in connection with the height above the crest, or lip of spill-way, and the slopes, would give at the level of the lip, a thickness at least equal to the greatest depth of water in the reservoir.

For the slopes, much depends upon the material used for embankments. If a good quality of earth is to be found, such as a sandy loam, containing a good proportion of clay and few stones, great thickness of embankment is not necessary. If the quality of the stuff is inferior, it must be made up for in quantity. Generally slopes of from $2\frac{1}{2}$ to $3\frac{1}{2}$ to 1 for the inside, and $2\frac{1}{2}$ to 1 for the outside are found sufficient. An excellent profile is obtained, with good material, by starting from the top with a slope of $2\frac{1}{2}$ to 1 on both sides, and at about half-way down introducing a level berme on the inside, from which a slope of 3 or $3\frac{1}{2}$ to 1 is continued down to the foot of the embankment, the outside slope being kept at the same slope of $2\frac{1}{2}$ throughout.

The manner in which the embankment

is formed, is a matter of great importance. It is necessary to put it in in such a manner as to minimize the subsequent settlement. It must not, therefore, be carried along like a railroad embankment, but brought up in horizontal layers from the bottom. Generally speaking, if the material be brought on in carts and wagons, and kept constantly moist by sprinkling, the travel of the vehicles and animals will be sufficient to secure the necessary degree of compactness. The face of the interior slope should be well pitched, or rip-rapped, as recommended by Mr. Jacob, and the exterior slope sodded, or sown to grass.

To return to the center wall, in order to say a few words as to the manner in which it, and in general all the masonry work of the dam, should be executed. The first essential of all hydraulic masonry is, that it shall be as nearly as possible water-tight. In order to effect this, there must be no vacancies whatever between any two contiguous stones, but the wall must be so perfectly laid that whatever is not stone, is mortar, and *compact* mortar. In order to secure perfect work, the stones used must be clean and bright, with sharp quarry faces. All soiled and dirty stones should be, if possible, discarded, and if not, they should be thoroughly cleaned, by brooms or brushes, or by washing, before using, for the mortar will not properly adhere if there be a loose coating of any foreign substance upon the surface of the stone. If necessary, they should be wet before laying, but frequently, particularly when Portland cement is used, the mortar throws off so much water that the chief difficulty is to keep the work sufficiently dry. When a piece of wall is to be worked upon, it should be carefully cleared of all loose and dry mortar; indeed, it is often necessary to keep men employed constantly in sweeping off the wall in order not to delay the masons in cleaning it up. This point should always be insisted upon, and in summer there should be a number of large watering-pots on the work, to sprinkle the surface after it has been swept off. Great care should also be taken to have the mortar properly mixed—a point too often neglected. The sand and cement should be thoroughly mixed together when dry, by repeated turnings, until the mass has become perfectly homogeneous, and of an uniform color.

It is generally a provision of the specifications, that stones must not exceed a certain limit in size—perhaps 8 or 10 cubic feet—it being supposed that small stones will be more easily bedded than large ones. I think that this is at least doubtful, and that stones measuring a cubic yard, or even a yard and a half, can be properly bedded with very little additional trouble. The gain in rapidity of execution, when larger stones are used, is very great, as a derrick can swing a large stone as quickly as a small one, and therefore many more yards can be laid in a day. Large stones also bed themselves more perfectly, from the greater force with which they compress the mortar under them, and drive out the superfluous water. Of course, stones should not be allowed of such dimensions as to run through the wall from front to rear, and it is always well to reserve the right to discard stones above, say, 10 cubic feet, by a clause to that effect, in the specifications.

After a stone has been set, it should be tested with bars, so as to see if it is entirely clear of everything, and swims on its bed, without rocking. No two stones, however small, should be allowed to touch each other without the intervention of compact mortar. When there is any doubt as to whether a stone is properly bedded, it should be raised, so as to see if the mortar under it has taken a good impression. It is frequently necessary to raise a stone a second and even a third time, before it can be pronounced to be satisfactorily set, particularly with inexperienced masons. The masons should, however, soon become accustomed to properly spreading the beds, so that a majority of the stones can be set by first intention. All work should be done with cranes or derricks, so that stones can be dropped in their places, and readily lifted again, when required. All spalls and small stones should be settled in their beds by a blow with the hammer. This scrupulous care in laying the masonry does not consume as much time as might be anticipated. There should be no trouble, with good and shapely stones, in laying from 30 to 35 cubic yards per day, per derrick (steam), with say six masons and a sufficient number of helpers, and yet take all the precautions—and many more—above indicated. Of course, to effect this, the work must be well systematized, and the derricks constantly fed with material, so that the masons are never kept waiting.

In putting down foundations, care must be taken to provide abundant pumping power. All pumping sumpts should be put down well below the bottom of the trenches they are designed to drain, otherwise these latter will never be kept properly dry. The sumpts should also, if possible, be placed entirely outside the lines of the masonry, so as to be out of the way, and not draw the material from under work already laid.

The management of the water is always one of the greatest difficulties connected with this class of work. The foundations extend entirely across the valley of the stream, and provision must be made for the passage of the water while the work is in progress. Very frequently, particularly in the case of small streams, the difficulty can be met by first building the gate chamber and setting the pipes and gates, and then turning the stream through them, thus getting rid of it at once. In larger streams this is not always practicable, as we may not be sure that the pipes will carry all the water in case of a freshet. In such cases, we may be obliged to omit setting the pipes till after the completion of the work, allowing the water to pass through the gate chamber and gallery in which they are subsequently to be set. When this course is pursued, the opening in the center wall must be so planned and built that it can be quickly and perfectly closed, when the time comes, by cut-stone pieces carefully dressed to fit. When the dam is otherwise completed, the stop-plank can be placed, and the water held back while the pipes and gates are set. In cases where the stream is so large that even these means are not sufficient, the problem becomes exceedingly difficult, and demands very careful special study.

Mr. Jacob dwells much upon the danger caused by the existence of springs underneath the embankment, and recommends that they be "taken up and carried away in proper drains sufficiently and securely puddled." Now, unfortunately, this is easier said than done. The site of an embankment is frequently covered by a multitude of small springs, the sources of which it is impossible to locate, and to which they cannot be traced. Besides, where should we carry them to? Probably the best way of dealing with such springs, is to get the site of the inner embankment well stripped down to the live earth, and then commence placing the embankment next to the center wall (not, of course, carrying it up so high as to create a *dump*, such as we have already seen should be avoided,) and then advance it out from the wall, toward the toe of the slope, in the hopes of smothering down the springs, and driving them, to some extent at least, back from under the embankment. Indeed, it is difficult to see what else can be done, and it may be fairly doubted if the existence of such springs is fraught with such serious menace to the work as Mr. Jacob mentions. Very large springs might occasion trouble, if the weight and compactness of the bank do not force them back, and, *if possible*, such should be traced to their source, and diverted within the reservoir. Unfortunately, these large veins of water are frequently fed by a number of separate springs, which finally run together, and if the flow be stopped in one point, it breaks out in another.

Large springs, such as have been just

mentioned, often give great trouble in closing the gap in the foundation of the center wall, which has been left for them to flow through temporarily, the difficulty being of the same class, though less in degree, as that occasioned by the main stream itself. They may sometimes be passed over from side to side, until a point is reached, above which they do not rise in any considerable volume. Sometimes they are best handled by reducing the width of the gap progressively, until it becomes a very narrow passage, and then closing it boldly, working in the water with mortar made of neat cement, without any admixture of sand. There can be no rule laid down for the treatment of such cases, which must be handled as best we may, the only thing necessary and sufficient being to get the wall so built that the water does not, finally, wash out the mortar and run through it. If we fail in accomplishing this in one way, there is nothing for it but to take out the defective part and try some other means, never leaving the job till the essential object aimed at has been accomplished.

In conclusion, it may be said that the building of dams and reservoirs belongs to one of the most important and responsible classes of work undertaken by the engineer, and no pains should be spared in their construction to ensure satisfactory results. It should be borne in mind that the first essential in a dam is stability; the second impermeability. At the present day, with all past experience as a guide, there should be no uncertainty as to the realization of the first of these essentials; the perfect fulfillment of the second cannot always be so surely counted upon, because it depends not only upon the quality of our work—which is a controllable factor—but also upon the character of the ground itself on which the work is placed, which may present disadvantageous features that we cannot wholly overcome. But by care and the exercise of sound judgment these adverse conditions can generally be modified to the extent of securing satisfactory, if not perfect results.

www.ingramcontent.com/pod-product-compliance
Lightning Source LLC
Chambersburg PA
CBHW030400170426
43202CB00010B/1433